500
AP Environmental Science Questions
to know by test day

Also in the 5 Steps series:
5 Steps to a 5: AP Environmental Science

Also in the 500 AP Questions to Know by Test Day series:
5 Steps to a 5: 500 AP Biology Questions to Know by Test Day
5 Steps to a 5: 500 AP Calculus Questions to Know by Test Day
5 Steps to a 5: 500 AP Chemistry Questions to Know by Test Day
5 Steps to a 5: 500 AP English Language Questions to Know by Test Day
5 Steps to a 5: 500 AP English Literature Questions to Know by Test Day
5 Steps to a 5: 500 AP Environmental Science Questions to Know by Test Day
5 Steps to a 5: 500 AP European History Questions to Know by Test Day
5 Steps to a 5: 500 AP Human Geography Questions to Know by Test Day
5 Steps to a 5: 500 AP Microeconomics/Macroeconomics Questions to Know by Test Day
5 Steps to a 5: 500 AP Physics Questions to Know by Test Day
5 Steps to a 5: 500 AP Psychology Questions to Know by Test Day
5 Steps to a 5: 500 AP Statistics Questions to Know by Test Day
5 Steps to a 5: 500 AP U.S. Government & Politics Questions to Know by Test Day
5 Steps to a 5: 500 AP U.S. History Questions to Know by Test Day
5 Steps to a 5: 500 AP World History Questions to Know by Test Day

5 STEPS TO A 5™

500 AP Environmental Science Questions to know by test day

Chris Womack, Jane P. Gardner, and
Stephanie Richards

New York Chicago San Francisco Lisbon London Madrid Mexico City
Milan New Delhi San Juan Seoul Singapore Sydney Toronto

The ***McGraw-Hill*** *Companies*

CHRIS WOMACK has been writing and teaching in the biological sciences, biotechnology, and health care for 15 years. A freelance writer and editor based in Austin, Texas, he holds a bachelor's degree in biology and a master's degree in journalism.

JANE P. GARDNER has master's degrees in geology and secondary education. She taught high school environmental science for four years before becoming a full-time science writer.

STEPHANIE RICHARDS teaches middle school and high school science at the Army and Navy Academy in Carlsbad, California. She holds a master's degree in education and a bachelor of science from UC San Diego.

1 2 3 4 5 6 7 8 9 10 11 12 13 14 15 QFR/QFR 1 9 8 7 6 5 4 3 2 1

ISBN 978-0-07-178074-2
MHID 0-07-178074-2

e-ISBN 978-0-07-178075-9
e-MHID 0-07-178075-0

Library of Congress Control Number 2011936617

Series interior design by Jane Tennenbaum

This book is printed on acid-free paper.

CONTENTS

INTRODUCTION

Congratulations! You've taken a big step toward AP success by purchasing *5 Steps to a 5: 500 AP Environmental Science Questions to Know by Test Day*. We are here to help you take the next step and score high on your AP exam so you can earn college credits and get into the college or university of your choice.

This book gives you 500 AP-style multiple-choice questions that cover all the most essential course material. Each question has a detailed answer explanation. These questions will give you valuable independent practice to supplement your regular textbook and the groundwork you are already doing in your AP classroom. This and the other books in this series are written by expert AP teachers who know your exam inside out and can familiarize you with the exam format, as well as questions that are most likely to appear on the exam.

You might be the kind of student who takes several AP courses and needs to study extra questions a few weeks before the exam for a final review. Or you might be the kind of student who puts off preparing until the last weeks before the exam. No matter what your preparation style is, you will surely benefit from reviewing these 500 questions, which closely parallel the content, format, and degree of difficulty of the questions on the actual AP exam. These questions and their answer explanations are the ideal last-minute study tool for those final few weeks before the test.

Remember the old saying "Practice makes perfect." If you practice with all the questions and answers in this book, we are certain you will build the skills and confidence needed to do great on the exam. Good luck!

—Editors of McGraw-Hill Education

Earth Science

1. Of the earth's three major layers, which one contains the greatest portion of molten rock?

(A) lithosphere
(B) core
(C) mantle
(D) crust
(E) hydrosphere

2. Which of the following forces helps keep rock in the earth's interior hot and molten?

(A) radioactivity
(B) global warming
(C) subduction
(D) solar radiation
(E) friction between tectonic plates

3. In the earth's interior, hot, molten rock rises, and cooler rock sinks in a process known as

(A) subduction
(B) erosion
(C) convergence
(D) convection
(E) the law of superposition

4. At a transform fault, two tectonic plates move

(A) alongside one another and in the same direction
(B) directly toward one another
(C) directly away from one another
(D) alongside each other and in opposite directions
(E) simultaneously down into the earth's interior

5. Which of the following examples orders geological time units from shortest to longest?

(A) epoch, eon, period, era
(B) period, era, eon, epoch
(C) epoch, period, era, eon
(D) eon, era, epoch, period
(E) era, eon, period, epoch

6. Based on his estimates of the time it takes rocks to form, James Hutton realized by the late 18th century that most geological changes occur as a result of

(A) volcanoes, earthquakes, hurricanes, and other catastrophic events over thousands of years
(B) sedimentation, heat, pressure, and other gradual processes over millions of years
(C) meteorites, comets, and asteroids falling from space over millions of years
(D) gravitational attraction from the sun, moon, and planets over thousands of years
(E) plants, animals, and microorganisms wearing and reshaping the surface over thousands of years

7. Magma fills the gap between tectonic plates, forming new crust at

(A) subsidence zones
(B) subduction zones
(C) sedimentation zones
(D) convergent faults
(E) divergent faults

8. Earthquakes release huge amounts of energy accumulated through which of the following processes?

(A) tectonic plates constantly pushing against one another
(B) compressed magma being forced to the surface through cracks and vents
(C) the relentless back-and-forth movement of ocean tides
(D) sudden heating of the earth's core
(E) disturbance of the earth's magnetic field by the moon's orbit

9. Which three processes are needed in order to convert metamorphic or igneous rock into sedimentary rock?

(A) heating, pressure, and stress
(B) erosion, heating, and transportation
(C) pressure, erosion, and heat
(D) cooling, pressure, and weathering
(E) weathering, erosion, and deposition

10. Using the Richter scale, an earthquake of magnitude 5.0 is how many times more powerful than an earthquake of magnitude 4.0?

(A) 1.25
(B) 5
(C) 10
(D) 50
(E) 125

11. Volcanoes tend to occur nearest to

(A) the centers of tectonic plates
(B) the highest points along the equator
(C) ancient meteor impact sites
(D) freshwater coastlines
(E) the boundaries of tectonic plates

12. Subduction zones tend to produce deep undersea trenches as a result of

(A) two tectonic plates moving directly away from each other
(B) two tectonic plates sliding past each other
(C) one tectonic plate being pulled apart at the center
(D) one tectonic plate being forced to bend under another
(E) two tectonic plates forcing each other to bend down into the earth

13. The earth has about how many tectonic plates?

(A) 5
(B) 15
(C) 55
(D) 105
(E) 155

14. Steno's "principle of original horizontality" helps geologists do which of the following?

(A) understand the angle at which a layer of sedimentary rock will form on an already inclined rock layer
(B) determine the relative ages of rock layers based on their order of deposition
(C) understand how waterborne sediments are deposited near the edge of a body of water
(D) predict volcano eruptions by indicating pools of high-pressure subsurface magma
(E) determine the thinnest portions of a tectonic plate using sound waves

15. Which of the following terms describes the exact location where an earthquake occurred?

(A) fault line
(B) epicenter
(C) subduction zone
(D) hypocenter
(E) hot spot

16. During an earthquake, Rayleigh waves produce

(A) a bobbing, up-and-down motion on the surface
(B) a shearing, side-to-side motion in the earth's interior
(C) serial compressions and expansions through the earth's interior
(D) a shearing, side-to-side motion on the surface
(E) recurring echoes through the earth's interior

17. Compared to the Cambrian era, the Precambrian era was

(A) about the same length
(B) about $1/25$ as long
(C) about twice as long
(D) more than 1,000 times longer
(E) about six times longer

18. Tsunamis, like the one that hit northeastern Japan in March 2011, are usually caused by

(A) magma suddenly released into the ocean
(B) earthquakes under the ocean floor
(C) rapid closing of an ocean trench
(D) subsidence of land near the ocean
(E) sudden changes in magma convection currents

19. If a nearby city experiences earthquake damage of intensity XI, the earthquake's Richter scale magnitude is most likely to be

(A) 0.9 or less
(B) 1.0 to 2.9
(C) 3.0 to 4.9
(D) 5.0 to 6.9
(E) 7.0 or greater

20. Which of the following clues can tell seismologists where an earthquake is likely to occur?

(A) changes in nearby magnetic fields
(B) tornadoes and other unusual weather patterns
(C) the phase of the moon
(D) historical earthquake records
(E) a volcanic eruption from a hot spot in the middle of a tectonic plate

21. According to Nicolaus Steno, if rock layer 1 sits on top of rock layer 2, the age of rock layer 1 is probably

(A) younger
(B) older
(C) the same
(D) impossible to estimate without more information
(E) impossible to estimate with any information

22. Nicolaus Steno's law of horizontality says that sediments get thinner at

(A) the center of a body of water
(B) the edge of a body of water
(C) the center of a tectonic plate
(D) the equator
(E) subduction zones

23. Most divergent plate boundaries are difficult for scientists to study because most are located

(A) at high elevations
(B) on the ocean floor
(C) in Antarctica
(D) deep in the earth's mantle
(E) in highly active earthquake zones

24. A fault intersects an imaginary horizontal plane in a line that points directly northwest. This line is an example of a

(A) fault trend
(B) transform fault
(C) fault strike
(D) fault plane
(E) convergent boundary

25. A well-developed and large collection of fractures along the edge of a tectonic plate is generally known as a

(A) fault plane
(B) fault trend
(C) fault zone
(D) fault line
(E) surface trace

26. The Hawaiian Islands and some other island chains result from a series of volcanoes occurring as a tectonic plate moves over a

(A) caldera
(B) subduction zone
(C) divergent boundary
(D) transform fault
(E) hot spot

27. Volcanoes can spew several types of gas, but the two main components are

I. oxygen gas
II. carbon dioxide
III. hydrogen gas
IV. hydrogen sulfide
V. water vapor

(A) I and II
(B) II and III
(C) III and IV
(D) IV and V
(E) II and V

28. A volcano that has not recently been active but may reactivate soon is known as

(A) extinct
(B) active
(C) dormant
(D) pyroclastic
(E) a hot spot

29. Which of the following is an example of an earthquake's primary effect?

(A) a fire
(B) a shaking movement
(C) a tsunami
(D) a flood
(E) a rockslide

30. Gases from a volcano can be very dangerous due to their

(A) velocity
(B) radioactivity
(C) temperature
(D) acidity
(E) mass

31. Nicolaus Steno developed his three laws for determining the relative age of geological samples in 1669. These are the law of superposition, the principle of original horizontality, and the principle of lateral continuity. In 1785, James Hutton introduced his principle of uniformitarianism, which establishes the earth's geological changes as very gradual and its age as much greater than several thousand years, which scientists previously believed.

(A) Considering sedimentary rock layers, how do Steno's first two laws support uniformitarianism's concept of an ancient earth?
(B) How does Steno's third law, the principle of lateral continuity, explain why some sedimentary layers are not contiguous and why they get very thin near their edges?
(C) The Himalayan range is an average of 8,848 meters high, and it currently rises by about 5 millimeters each year, as the Indo-Australian tectonic plate collides with the Eurasian plate, but erosion keeps its height in check. Assuming that the Himalayas started at sea level when the continents themselves first met 10 million years ago, what is the average rate at which the mountain range rose in that time, accounting for the effects of erosion?
(D) With the same assumption that the Himalayas started at sea level, how fast would the Himalayas be rising today if the continents collided only 6,000 years ago?

32. The Pacific Ring of Fire is the name of a tectonically active area around the border of almost the entire Pacific Ocean. The volcanoes in the Ring of Fire result from the Pacific plate's movement toward or away from several different continental plates, including the North American plate, the Indo-Australian plate, and the Antarctic plate. Japan's calamitous March 2011 earthquake and tsunami were results of tectonic activity along nearby portions of the Ring of Fire. Japan's iconic Mount Fuji is also a product of the ring—it's the country's highest volcano. Japan lies on a complex intersection of the Pacific plate, the Philippine plate, the Eurasian plate, and the North American plate. Japan and other islands in the area form a line that tends to follow local plate boundaries, and the region is very rich in volcanoes.

(A) What causes the Pacific plate to move?
(B) What process generates magma circulation?
(C) Which direction is the Pacific plate probably moving?
(D) The Nazca plate lies under the Pacific Ocean between the Pacific plate and the South American plate. Between the Nazca plate and the Pacific plate lie a series of ocean ridges that form a jagged line, rather than a straight line, on the ocean floor. What is this called, and what lies at each offset part of the ridge?

33. To the west and southwest of Wyoming's Yellowstone National Park, there is evidence of several volcanic eruptions that have occurred over the past 17 million years. Today, the park has many features associated with volcanically active areas, including hot springs and geysers. But Yellowstone lies well within the North American plate. There are huge, mostly flat regions in the area around Yellowstone, where mountains once stood. These areas formed during huge volcanic eruptions, and they are oval or nearly circular in shape.

(A) What is the likely cause of volcanoes in the Yellowstone region over the past 17 million years?
(B) What direction is the North American plate moving?
(C) What is the name of the large flat areas left by ancient volcanoes, and what causes them?

Atmospheric Conditions

34. Currently, the atmosphere surrounding earth is composed mainly of

(A) nitrogen
(B) oxygen
(C) carbon dioxide
(D) helium
(E) methane

35. The Coriolis effect

(A) changes the circulation direction of water running down a drain
(B) changes the direction of river flow
(C) changes the direction of winds all around earth
(D) causes winds in the Northern Hemisphere to move to the east
(E) causes winds in the Southern Hemisphere to move to the west

36. Temperatures within the stratosphere increase with elevation due to the presence of

(A) the ozone layer
(B) numerous clouds
(C) high-speed winds
(D) temperature inversion
(E) electrically charged ions

37. Which two factors allow earth to experience four distinct seasons?

(A) latitude, solar intensity
(B) tilt of axis, rotation on axis
(C) tilt of axis, revolution around sun
(D) distance from sun, revolution around sun
(E) distance from sun, place in orbit around sun

38. Evaporation involves the conversion of

(A) latent heat into solar energy
(B) solar energy into latent heat
(C) chemical energy into solar energy
(D) potential energy into kinetic energy
(E) chemical energy into mechanical energy

39. The Coriolis effect is the result of

(A) earth's gravitational pull
(B) earth's rotation on its axis
(C) the gravitational pull of the moon
(D) the revolution of earth around the sun
(E) the differential heating of earth's surface

40. An isobar on a weather map connects areas of

(A) low pressure
(B) high pressure
(C) equal elevation
(D) equal temperature
(E) equal atmospheric pressure

41. Which is the correct sequence of the layers of the earth's atmosphere from the earth's surface toward space?

(A) stratosphere, troposphere, mesosphere, thermosphere
(B) troposphere, mesosphere, stratosphere, thermosphere
(C) mesosphere, stratosphere, thermosphere, troposphere
(D) troposphere, stratosphere, mesosphere, thermosphere
(E) thermosphere, stratosphere, mesosphere, troposphere

42. Precipitation will fall when the humidity is at

(A) 10%
(B) 25%
(C) 50%
(D) 75%
(E) 100%

43. Relative humidity is a measurement of

(A) temperature of the air and elevation
(B) elevation and the amount of water vapor in the air
(C) the amount of water vapor the air can hold and latitude
(D) air temperature and latitude
(E) air temperature and the amount of water vapor in the air

44. Temperature inversions within the troposphere are likely to occur in association with

(A) oceans
(B) deserts
(C) urban areas
(D) mountain ranges
(E) the equator

45. Temperatures on earth are influenced by all of the following EXCEPT

(A) elevation
(B) latitude
(C) the position of the moon
(D) proximity to bodies of water
(E) ocean currents

46. Trade winds in the Northern and Southern Hemispheres deflect toward

(A) the equator
(B) the poles
(C) continental land masses
(D) the temperate zones
(E) higher latitudes

47. Atmospheric pressure increases with

(A) increasing altitude
(B) decreasing altitude
(C) increasing temperature
(D) longitude
(E) latitude

48. Hurricanes begin with

(A) a drop in temperature
(B) periods of wind shear
(C) a drop in barometric pressure
(D) incoming tides
(E) a rise in barometric pressure

49. Which of the following is a characteristic of the troposphere?

(A) part of the atmosphere that contains the ozone layer
(B) an increase of temperature with elevation
(C) the aurora borealis
(D) layer where our weather occurs
(E) cruising altitude for commercial aircraft

50. A tropical storm is renamed a hurricane when wind speeds exceed

(A) 25 knots, or 46 km/hr
(B) 57 knots, or 106 km/hr
(C) 64 knots, or 119 km/hr
(D) 73 knots, or 135 km/hr
(E) 80 knots, or 148 km/hr

51. On average, an ENSO event occurs every

(A) 1 to 2 years
(B) 2 to 7 years
(C) 7 to 10 years
(D) 15 to 25 years
(E) 30 to 40 years

52. Which of the following is characteristic of an El Niño year?

(A) Warm water in the Pacific is pushed westward.
(B) Cold water upwells along the South American coast.
(C) Warm water flows eastward, reaching South America.
(D) Cold water in the Pacific is pushed westward.
(E) Warm water upwells along the South American coast.

53. What is the difference between an El Niño event and a La Niña event?

(A) La Niña involves cool Pacific Ocean surface temperatures, while El Niño involves warmer ocean surface temperatures.
(B) El Niño is a milder version of La Niña.
(C) La Niña events occur in the Pacific Ocean, and El Niño events occur in the Atlantic Ocean.
(D) La Niña events occur in the Atlantic Ocean, and El Niño events occur in the Pacific Ocean.
(E) La Niña events increase temperatures worldwide, and El Niño events decrease temperatures.

54. The global impact of an ENSO includes all of the following EXCEPT

(A) drought in the western Pacific
(B) flooding in the eastern Pacific
(C) disruption of the fish population along South America
(D) a decrease in the number of Atlantic hurricanes
(E) an increase in the number of Atlantic hurricanes

55. The lowest barometric pressure associated with a hurricane can be found

(A) in the eye
(B) in the eye wall
(C) along the outer edges of the storm
(D) when the hurricane is a tropical depression
(E) when the hurricane is a tropical storm

56. The fastest winds measured on earth were associated with a(n)

(A) hurricane
(B) tropical storm
(C) ENSO event
(D) tornado
(E) tropical depression

57. A severe tornado may have wind speeds between

(A) 40 and 72 mph
(B) 73 and 112 mph
(C) 113 and 157 mph
(D) 158 and 206 mph
(E) 207 and 260 mph

58. Jet streams are located within the

(A) troposphere
(B) stratosphere
(C) ozone layer
(D) mesosphere
(E) ionosphere

59. Sunspot activity directly affects the temperature within the

(A) thermosphere
(B) mesosphere
(C) ionosphere
(D) stratosphere
(E) troposphere

60. The ozone layer absorbs

(A) infrared radiation from the sun
(B) ultraviolet radiation from the sun
(C) positively charged ions
(D) heat radiated off earth's surface
(E) water vapor

61. Air pressure is measured using a

(A) thermometer
(B) anemometer
(C) barometer
(D) hydrometer
(E) wind vane

62. How does the amount of water warm air can hold compare to the amount of water colder air can hold?

(A) Warm air can hold more water than cold air.
(B) Warm air can hold less water than cold air.
(C) Warm air and cold air can hold the same amount of water.
(D) Warm air is not able to hold water, and cold air is.
(E) Cold air is not able to hold water, and warm air is.

63. Most Atlantic hurricanes form off the coast of

(A) South America
(B) the Philippines
(C) the southern United States
(D) Europe
(E) Africa

64. The ozone layer in the stratosphere is very important to life on earth, particularly organisms on land. It filters ultraviolet radiation from the sun, protecting organisms and keeping the temperature on earth at a tolerable level. Ozone in the troposphere is classified as a pollutant. It can adversely affect the health of humans and other organisms.

(A) What is the function of the ozone layer in the stratosphere?
(B) What chemical reactions play a part in the degradation of the ozone layer in the stratosphere?
(C) What is the source of the ozone in the troposphere? What health problems can it cause?

65. Most places on earth experience distinct seasons featuring differences in temperature and precipitation. Earth is not the only planet in the solar system with seasons, although the changes in seasons are marked by different phenomena elsewhere.

(A) What causes earth's seasons?
(B) The earth and the sun are closer during the Northern Hemisphere's winter than during its summer. Explain how this is possible.
(C) How do the seasons impact the climate of particular locations on earth?

66. The Atlantic hurricane season lasts from June 1 to November 30. Some climate models predict that hurricanes will become more common each year. Others indicate that the number of Atlantic hurricanes may fluctuate each year but remain fairly constant in the long term.

(A) Why is the Atlantic hurricane season limited to the period between June and November? What conditions can cause a hurricane to occur?

(B) How might climate change cause hurricanes to become more common?

(C) What evidence contradicts the view that climate change will cause more hurricanes?

Global Water Resources and Use

67. In response to the threats to our oceans, marine scientists signed a statement in 1998 explaining the problems to the public. This statement was called the

(A) Marine Conservation Act
(B) *Troubled Waters* statement
(C) Marine Conservation statement
(D) Environmental Impact statement
(E) none of the above

68. An influent stream is

(A) entirely above the water table
(B) a result of precipitation
(C) maintained by groundwater seeping in
(D) a perennial stream
(E) both A and B

69. Which of the following illustrates a water budget?

(A) water volume – surface area = total % of water
(B) stream flow – subsurface flow = evaporation
(C) precipitation – evaporation = runoff
(D) precipitation – consumptive use = water available
(E) influent stream + effluent stream = groundwater

70. The Kissimmee River in Florida is an example of

(A) flood control that led to habitat destruction
(B) drought and its impact on water supply
(C) groundwater overdraft and sinkholes
(D) desalination and declining turtle populations
(E) deforestation and water runoff

71. What water management technique is used to control floods and erosion, improve drainage, and improve navigation?

(A) catchwater system
(B) condensation
(C) overdraft
(D) channelization
(E) desalination

72. *Dendritic drainage* is

(A) percolation of water into groundwater
(B) a branching drainage pattern of streams
(C) seeping of water out of discharge zones
(D) evapotranspiration from plants
(E) the release of precipitation from clouds

73. When groundwater is removed, the soil sinks in a process known as

(A) saturation
(B) stabilization
(C) subsidence
(D) streambed channelization
(E) sustainability

74. The amount of moisture in the air compared with how much moisture the air could potentially hold is known as

(A) relative humidity
(B) absolute humidity
(C) precipitation
(D) evaporation potential
(E) none of the above

75. Hoover Dam and Glen Canyon Dam manage which river?

(A) Mississippi River
(B) Colorado River
(C) Kissimmee River
(D) River Ouse
(E) Potomac River

76. Coastal beaches and barrier islands are eroded by

(A) jetties
(B) percolation
(C) longshore currents
(D) compaction
(E) desalination

77. Vernal pools are

(A) mangrove swamps that are endangered
(B) forest bottomland that is flooded
(C) storm surges that are stagnant
(D) temporary pools of water
(E) bogs that serve as primary habitat

78. Water is stored an average of 8 to 10 days in which of the following reservoirs?

(A) rivers and streams
(B) soil moisture
(C) oceans
(D) snow and glaciers
(E) atmosphere

79. Land masses hold what percentage of the earth's water?

(A) 97%
(B) 3%
(C) 0.001%
(D) 72%
(E) 66%

80. Resources such as groundwater that are being consumed faster than they can be resupplied are said to be

(A) infiltrated
(B) highly salinized
(C) in the zone of saturation
(D) in a drainage basin
(E) nonrenewable

81. In general, ocean salinity

(A) increases with precipitation
(B) increases with depth
(C) decreases with depth
(D) is less than freshwater salinity
(E) is less than groundwater salinity

82. Ocean circulation is changed by

(A) wind
(B) differences in water density
(C) differences in salinity
(D) differences in temperature
(E) all of the above

83. Warm Caribbean water travels north past Canada to Northern Europe in the

(A) Equatorial Countercurrent
(B) Humboldt Current
(C) Labrador Current
(D) Gulf Stream Current
(E) Japan Current

84. The boundary between the zone of saturation and zone of aeration is known as the

(A) watershed
(B) water table
(C) aquifer recharge zone
(D) runoff
(E) zone of groundwater

85. The withdrawal rate of water is less than 1% of the annual supply in which of the following countries?

(A) Canada
(B) Brazil
(C) United States
(D) Israel
(E) both A and B

86. All the land from which water drains to a common lake or river is considered to be part of the same

(A) watershed
(B) aquifer
(C) runoff zone
(D) water table
(E) both A and C

87. Building and construction sites mainly affect water purity by

(A) producing loose sediment that washes away in heavy rainfall
(B) producing hazardous chemicals that infiltrate groundwater
(C) creating more evaporation of groundwater
(D) groundwater mining of local resources
(E) increasing the zone of saturation

88. Using reclaimed water, preventing runoff, and using dry cooling systems in industrial processes are all methods of

(A) precipitation
(B) conservation
(C) groundwater mining
(D) aquifer recharge
(E) irrigation control

89. Using excess groundwater in coastal regions causes

(A) a sustainable water supply
(B) condensation nuclei
(C) surface water to decline
(D) saltwater intrusion
(E) both C and D

90. A drought occurs when

(A) rainfall is 50% below average for a period of 10 days or more
(B) rainfall is 50% below average for a period of 21 days or more
(C) rainfall is 70% below average for a period of 21 days or more
(D) rainfall is 70% above average for a period of 21 days or more
(E) rainfall is 70% below average for a period of 10 days or more

91. An acre-foot of water is

(A) more expensive using conservation methods
(B) the number of square feet of water in an acre
(C) the amount of water needed by an acre of farmland
(D) an acre of water one foot deep
(E) none of the above

92. The draining of wetlands was authorized by which legislation?

(A) Watershed Protection and Flood Prevention Act
(B) Coastal Zone Management Act
(C) Coastal Barrier Resources Act
(D) National Wildlife Refuge System
(E) Wild and Scenic Rivers Act

93. The endangerment of native fish, diminished water quality in streams, and flooding of farmland are all potential consequences of

(A) sinkholes
(B) dams and reservoirs
(C) gray water systems
(D) subsidence
(E) groundwater overdraft

94. The Aral Sea provides an example of the consequences of

(A) dam removal on local populations
(B) desalination on energy costs
(C) diverting water for agricultural purposes
(D) water conservation for domestic use
(E) restoration of wetlands

95. Hydrology, types of vegetation, and types of soil are used to determine if an area is

(A) a wetlands area
(B) a drainage basin
(C) a perched water table
(D) an unconsolidated aquifer
(E) an ecological pyramid

96. In the United States, what percentage of the total annual renewable water supply is withdrawn for use by people?

(A) 60%
(B) 50%
(C) 70%
(D) 40%
(E) 80%

97. The hydrologic, or water, cycle can take many different paths in the transfer of water.

(A) Describe three different paths a drop of water can take starting from a rain cloud. Describe each path using terms such as *infiltration*, *evaporation*, and *transpiration*.
(B) Discuss one of these paths using terms such as *influent*, *effluent*, *perennial*, or *ephemeral* in a specific example (e.g., a stream coming from the mountains).
(C) The city of Springfield, Oregon, gets the majority of its water supply from groundwater. Explain what happens when too much of this water is extracted for use by the population.
(i) Describe potential effects on nearby plants and animals.
(ii) Describe how removing water can actually lead to flooding.
(iii) Describe what changes can happen to the city itself from groundwater overdraft.

98. In the wake of Hurricane Katrina, you have been appointed to a special committee by the government to look at ways to prevent flooding in other cities in the future.

(A) Discuss how both natural events and man-made changes contribute to flooding.
(B) Describe two methods of flood control that are currently in use in the United States and the drawbacks of each.
(C) The Watershed Protection and Flood Prevention Act of 1954 was intended to alleviate the problem of flooding where wetlands were located. Explain whether or not this legislation was effective and why or why not.
(D) Outline your recommendations to the government, using the concept of watershed management.

99. Glen Canyon Dam in Page, Arizona, is the second largest dam on the Colorado River. There is opposition to this dam and the building of others by many environmental groups. China now has the world's largest dam, and it, too, is criticized for its environmental impact.

(A) Describe three benefits that dams provide to people.

(B) The city of Springfield has plans to build a new dam. Assume that you work for an environmental group that opposes the dam. Using the example of the Three Gorges Dam in China, write a letter to be mailed to the residents explaining why they should reject funding for this project in the next election.

(C) Identify and describe four ways that the water supply can be managed to avoid building a dam in the future.

Soil and Soil Dynamics

100. Rocks that are deposited in layers are classified as

I. igneous
II. sedimentary
III. metamorphic

(A) I only
(B) II only
(C) III only
(D) I and II
(E) II and III

101. All of the following are steps in the process of diagenesis EXCEPT

(A) compaction
(B) cementation
(C) recrystallization
(D) melting
(E) chemical changes

102. Igneous rocks are grouped according to

(A) composition
(B) grain size
(C) color
(D) parent material
(E) shape

103. If the metamorphic rock phyllite is melted, it will likely become

(A) the sedimentary rock shale
(B) magma
(C) sediment
(D) the igneous rock granite
(E) the metamorphic rock gneiss

104. Which statement BEST describes physical weathering?

(A) the breaking of a bigger rock into smaller pieces with a change in chemical composition
(B) the breaking of a bigger rock into smaller pieces without a change in chemical composition
(C) the compaction and cementation of sediment into a sedimentary rock
(D) the gradual cooling of magma into an igneous rock
(E) the change of one rock into a metamorphic rock due to heat and pressure

105. The chemical weathering process of hydrolysis breaks down a rock that reacts with

(A) oxygen
(B) carbonic acid
(C) percarbonic acid
(D) water
(E) sulfuric acid

106. Solubility describes a mineral's

(A) ability to form oxides
(B) susceptibility to weathering with carbonic acid
(C) ability to dissolve in water
(D) reaction with environmental acids
(E) formation when water reacts with a different mineral

107. The greatest amount of organic matter in the soil is found in

(A) the bedrock
(B) the accumulation zone
(C) C horizon
(D) B horizon
(E) A horizon

108. Plants, mosses, and bacteria break down rocks through

(A) chemical weathering
(B) exfoliation
(C) unloading
(D) physical weathering
(E) both physical weathering and chemical weathering

109. The type of soil that contains iron oxides and aluminum-rich clays is called

(A) parent material
(B) pedocal
(C) pedalfer
(D) laterite
(E) bedrock

110. In which of the following environments are laterite soils more likely to form?

(A) grasslands
(B) tundra
(C) conifer forests
(D) tropics
(E) deciduous forests

Use Figure 4.1 to answer questions 111–113.

111. Which layer in the soil horizon serves as the parent material for the soil that forms?

(A) O horizon
(B) A horizon
(C) B horizon
(D) C horizon
(E) bedrock

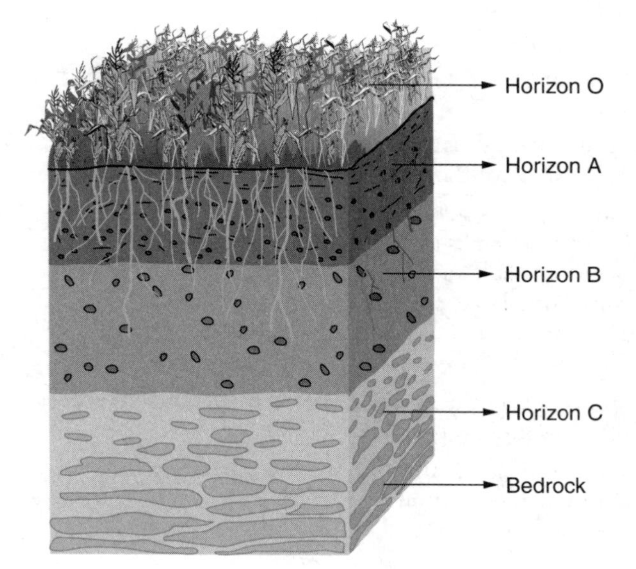

Figure 4.1

112. The topsoil is composed of

(A) O horizon
(B) A horizon
(C) B horizon
(D) O and A horizons
(E) A and B horizons

113. The B horizon is also sometimes referred to as the

(A) zone of accumulation
(B) zone of leaching
(C) parent materials
(D) topsoil
(E) humus

114. Pedocal soils are characterized by large accumulations of

(A) iron oxide
(B) calcium carbonate
(C) aluminum-rich clay
(D) organic material
(E) humus

115. Soil erosion has increased due to all of the following activities EXCEPT

(A) contour plowing
(B) logging
(C) development
(D) construction
(E) farming

116. One of the methods farmers use to prevent soil erosion is a practice called *terracing*. Which of the following best describes this practice?

(A) plowing fields perpendicular to the slope of hills
(B) leaving old stalks and other plant material on fields
(C) planting alfalfa or clover on fields
(D) converting large, steep fields into smaller fields
(E) planting different crops every year in a field

117. Erosion may occur due to the repeated freeze-and-thaw cycle in some climates. This process is called

(A) granular disintegration
(B) joint block separation
(C) frost wedging
(D) unloading
(E) salt wedging

118. Salt wedging is a significant cause of mechanical weathering in

(A) tropical ecosystems
(B) deciduous forests
(C) grasslands
(D) tundra ecosystems
(E) deserts

119. Dolomitization occurs when

(A) limestone is turned into dolomite
(B) dolomite is turned into marble
(C) marble is turned into dolomite
(D) sediment is turned into dolomite
(E) shale is turned into slate

120. Clastic sedimentary rocks are composed primarily of

(A) organic material
(B) felsic and mafic minerals
(C) weathered particles of sediment
(D) recrystallized minerals from solution
(E) evaporated materials such as halite or gypsum

121. Studying rocks through a microscope or with a hand lens is called

(A) stratigraphy
(B) lithology
(C) dissolution
(D) lithification
(E) diagenesis

122. A rock that forms deep in the crust will weather by the process of unloading as

(A) the rock is pushed deeper into the crust
(B) more sediment is deposited on top of the rock
(C) the rock is remelted into magma
(D) frost wedging occurs
(E) there is erosion of the rocks above

123. Trees, plants, and other organisms such as worms can weather a rock by

(A) chemical weathering
(B) physical weathering
(C) both chemical and physical weathering
(D) unloading
(E) hydrolysis

124. In what type of environment is frost wedging prevalent?

(A) cooler, wet climates
(B) warm, wet climates
(C) warm, dry climates
(D) cold, dry climates
(E) equatorial climates

125. Rainwater mixes with which of the following to form carbonic acid?

(A) hydrochloric acid
(B) carbon dioxide
(C) carbon monoxide
(D) sulfur dioxide
(E) methane

126. What is a benefit of contour plowing?

(A) an increase in soil fertility
(B) a slowdown in runoff
(C) a decrease in windblown soil
(D) water forms deep pools
(E) cultivation on a steep slope is possible

127. The oxidation of a rock may produce

(A) H_2O
(B) Fe_2O_3
(C) H_2CO_3
(D) H_2SO_3
(E) CO_2

128. A rock composed of pebble-sized sediment with edges that are not well rounded is called

(A) sandstone
(B) shale
(C) conglomerate
(D) breccia
(E) siltstone

129. If the igneous rock granite is subjected to intense heat and pressure but does not melt, it will most likely become

(A) sandstone
(B) magma
(C) lava
(D) gneiss
(E) basalt

130. Current farming methods typically use one of two approaches: conventional or sustainable agriculture.

(A) How do conventional agriculture and sustainable agriculture differ in their approaches to controlling weeds, suppressing insect pests, and adding nutrients to the soil?
(B) Describe and explain three ways that conventional agriculture disrupts the natural ecology of an area.
(C) Describe three sustainable agriculture techniques that avoid disruption of land.

131. Physical and chemical weathering play a significant role in the rock cycle and in the formation of sedimentary rocks.

(A) Describe three methods of physical weathering and three methods of chemical weathering that can change a rock at or near earth's surface.
(B) After a rock has been weathered and becomes sediment, describe how it might become a sedimentary rock.
(C) How does sediment change as it moves through the rock cycle?

132. There are three types of rock: igneous, sedimentary, and metamorphic.

(A) Describe the formation of igneous rocks using the specific examples of basalt and granite.
(B) Sandstone and shale are two sedimentary rocks. Use these two examples to describe how sedimentary rocks form.
(C) Describe the formation of metamorphic rocks using the specific examples of gneiss and marble.

Ecosystem Structure and Diversity

133. Which of the following ecological levels of organization are arranged from largest to smallest?

(A) ecosystem, population, community
(B) biosphere, ecosystem, community
(C) community, population, ecosystem
(D) population, community, biosphere
(E) biosphere, population, ecosystem

134. Compared with a population with low genetic diversity, a genetically diverse population is more likely to include individuals that

(A) are distributed over a large range
(B) vary greatly by age
(C) vary greatly by appearance and behavior
(D) each carry a large number of genes
(E) produce a large number of offspring

135. The portion of earth's biosphere that is closest to the core is the

(A) crust
(B) hydrosphere
(C) asthenosphere
(D) mantle
(E) atmosphere

136. Which of the following is MOST likely to give an individual a larger range of tolerance than other members of its population?

(A) spending more time in extremely hot and cold conditions
(B) having a relatively large number of offspring
(C) genetic similarity to the rest of the population
(D) genetic differences from the rest of the population
(E) spending more time in constant chemical conditions

137. Genetic diversity, species diversity, ecological diversity, and functional diversity are best characterized as elements of

(A) cultural diversity
(B) trophic levels
(C) abiotic resources
(D) biological diversity
(E) biospheres

138. Ecosystems typically include all of the following EXCEPT

(A) decomposers
(B) producers
(C) abiotic chemicals
(D) consumers
(E) speculators

139. A secondary consumer obtains X energy from eating a primary consumer with a total energy of Y. What is the relationship of X and Y?

(A) $\frac{1}{x} = Y$
(B) $X = Y$
(C) $X < Y$
(D) $X > Y$
(E) $2Y = X$

140. Which of the following types of organisms are usually the first to suffer when an ecosystem is disrupted?

(A) decomposers and detritus feeders
(B) producers
(C) primary consumers
(D) secondary consumers
(E) tertiary consumers

141. Of the following pairs of ecosystems, which has the highest net primary productivity?

(A) swamps and estuaries
(B) woodland and agricultural land
(C) savannas and coniferous forests
(D) open ocean and tundra
(E) freshwater lakes and continental shelves

142. All of the following help determine an animal's niche EXCEPT

(A) its temperature-range tolerance
(B) physical damage it has sustained
(C) its adaptive traits
(D) its typical territorial size
(E) species that act as its predators and prey

143. When biologists speak of "survival of the fittest," which of the following attributes do they consider most important to the concept?

(A) the ability of the fastest individuals to avoid predators
(B) the ability of the strongest individuals to capture prey
(C) the ability of the most fertile individuals to successfully reproduce
(D) the ability of the most high-tolerance individuals to withstand temperature extremes
(E) the ability of the most colorful individuals to attract mates

144. Only 5% of a population of lizards survive a major flood, and by chance, the majority of survivors are green. If the original population was composed mostly of brown individuals, then which phenomenon has likely occurred?

(A) evolutionary divergence
(B) a bottleneck event
(C) parallel evolution
(D) natural selection
(E) differential reproduction

145. Compared to beneficial mutations, harmful mutations are

(A) more common
(B) equally common
(C) nonexistent
(D) less common
(E) extremely rare

146. In which of the following situations has evolution most likely taken place?

(A) An individual produces only taller offspring.
(B) A population produces offspring that are representative of the gene pool.
(C) A population recombines with an isolated group of distantly related individuals.
(D) A population's allele frequencies have changed over time.
(E) An individual reproduces more successfully than others in its population.

147. After splitting into two isolated populations, genetic changes make it impossible for an individual from one of the two populations to reproduce with an individual from the other. Which term most accurately describes what has happened?

(A) a bottleneck event
(B) differential reproduction
(C) natural selection
(D) speciation
(E) coevolution

148. A founder group of cats arrives on an isolated island and after thousands of generations produces several different cat species that each prey on different kinds and sizes of animals. This is an example of

(A) evolutionary divergence
(B) coevolution
(C) commensalism
(D) competitive exclusion
(E) secondary succession

149. Which of the following best describes the term "endemic species"?

(A) a species colonizing a far-flung new range
(B) a species that benefits from its relationship with a second species
(C) a species found in one specific region
(D) a species that spreads disease
(E) a species whose genetic diversity is harmed by a catastrophe

150. All of the following are elements of an organism's habitat EXCEPT its

(A) food
(B) water
(C) shelter
(D) living space
(E) reproductive strategy

151. A species is most likely to go extinct in a changing environment if it has

(A) high genetic diversity
(B) a large niche
(C) a narrow niche
(D) recent evolutionary adaptations
(E) evolutionary divergence from a founder population

152. One species benefits from its relationship with another species for whom the relationship does no harm nor good. This relationship is best described as

(A) commensalism
(B) parasitism
(C) ecological succession
(D) mutualism
(E) competitive exclusion

153. As a series of communities of organisms colonize a new area, trees become established

(A) early in secondary succession
(B) early in primary succession
(C) late in primary succession
(D) late in secondary succession
(E) early in tertiary succession

154. During primary succession, which of the following organisms often serve as pioneer species?

(A) lichens
(B) grasses
(C) small herbs
(D) trees
(E) bushes

155. Southeast Asian flying foxes are responsible for pollinating most durian tree flowers and spreading durian seeds. Durian fruit and durian trees support a vast array of other species, such that a decline in the flying fox population could have disastrous effects for the local ecology. As a result, flying foxes can probably be considered

(A) an indicator species
(B) an invasive species
(C) a keystone species
(D) a parasitic species
(E) a pioneer species

156. Coastal wetlands are ecologically important because they do all of the following EXCEPT

(A) reduce the impact of storms coming inland
(B) harbor a high level of biodiversity
(C) help to cycle nutrients
(D) provide important habitats for both marine and terrestrial species
(E) offer an important refuge for large marine mammals

157. Ecologists divide the open ocean into the euphotic zone, the bathyal zone, and the abyssal zone based mostly on

(A) the creatures that live in each
(B) the amount of sunlight that each receives
(C) the salinity of each
(D) the temperature of each
(E) the ratio of producers to consumers in each

158. In which freshwater lake zone would you expect to find the greatest concentration of waterborne plants?

(A) the benthic zone
(B) the profundal zone
(C) the limnetic zone
(D) the littoral zone
(E) none of the above

159. Conservation biologists have an emergency action plan for identifying and protecting about 25 "hot spots" because they contain

(A) the majority of the earth's freshwater reserves
(B) about $\frac{3}{5}$ of the earth's mineral nutrients
(C) about $\frac{2}{3}$ of the earth's terrestrial biodiversity
(D) about $\frac{1}{3}$ of the earth's marine biodiversity
(E) about $\frac{3}{5}$ of the earth's estuaries

160. Each of the following might be part of an ecological restoration project EXCEPT

(A) replanting a forest in a heavily logged area
(B) removing a dam to allow a river to run naturally
(C) reintroducing native species to a grassland
(D) removing invasive species from a wetland
(E) creating a wetland in a ruined grassland

161. Which South American country devotes a greater share of its land to biodiversity protection than any other country in the world?

(A) Brazil
(B) Costa Rica
(C) Nicaragua
(D) Venezuela
(E) Chile

162. Which of the following ocean zones contains the greatest total number of producers?

(A) the euphotic zone
(B) the bathyal zone
(C) the abyssal zone
(D) the coastal zone
(E) the estuarine zone

163. The silversword alliance is the name of a set of about 30 related species of plants endemic to Hawaii. Members of the sunflower family, the silversword alliance plants are all very closely related, having the same number of chromosomes and nearly the same genes. They probably developed from a single ancestor seed that came to the islands from North America about 5.2 million years ago. These species often feature long, thin leaves, sometimes with a green-silver color that gives them their collective name. But they show an astonishing morphological variety, from the small, bush-size *Argyroxiphium sandwicense* to the tall, palm tree–like *Wilkesia gymnoxiphium*. The tall, branching tree *Dubautia latifolia* lives only in the forests of the island of Kauai, while *Argyroxiphium sandwicense* recently lived only on steep cliffs on the island of Hawaii, although it once populated much of the upper portion of volcano Mauna Kea until herbivorous mammals came to the island. Recent efforts have repopulated some of the gravelly, nearly lifeless volcanic soil of Mauna Kea with *Argyroxiphium sandwicense*.

(A) Why does the silversword alliance lack genetic diversity despite the morphological diversity among its species?
(B) What is the connection between the silversword alliance's evolutionary divergence and its members' current habitats?
(C) Which of the above species is likely involved in primary succession, and why?
(D) Why did *Argyroxiphium sandwicense* probably not have some form of protection from herbivorous mammals?

164. The great-tailed grackle began to breed in Central Texas as late as the 1910s. A yellow-eyed relative of the crow measuring about 40 centimeters from the tip of its tail to the end of its long, thin beak, the great-tailed grackle settled the area after large towns and cities began to develop, and people cleared thick brush and forest for agriculture and widely spaced shade trees. Great-tailed males are black and iridescent blue, while females are smaller brown and yellow birds. Their relative, the 28-centimeter common grackle, once dominated Central Texas but is now only a visitor, even though both species have similar omnivorous diets, eating insects, small fish, berries, and unattended human food. Another related species, the boat-tailed grackle, still has essentially the same habitat as it had before large-scale human settlements came to Texas—it still lives along the Gulf of Mexico coast, overlapping with great-tailed territory in many places, although the boat-tailed bird dominates these areas.

(A) What is the likely reason that the common grackle left Central Texas when the great-tailed grackle moved in?
(B) In areas where the great-tailed and boat-tailed grackles both live, what is the boat-tailed grackle's realized niche?
(C) After the arrival of the great-tailed grackle, how does the boat-tailed grackle keep living in the same areas, while the common grackle did not?

165. Lichens are composed of two different species, a fungus and a photosynthetic microorganism—either a cyanobacterium or an alga, and sometimes both. Together they slowly grow on a surface as a plantlike mass that appears to be a single organism. They can grow in some of the most inhospitable conditions on earth, such as high elevations, deserts, bare rock, and frozen ground. The fungus provides protection from the elements and reduces water loss by covering the whole structure in tough filaments. Sometimes the fungus slowly dissolves the substrate that its lichen is growing upon and distributes minerals throughout the lichen. All lichen fungi help to gather water, and they help to gather minerals from dust and rain. Meanwhile, the photosynthetic microorganism converts carbon dioxide in the air to carbohydrates for food that it shares with the fungus. Neither the fungus, alga, nor cyanobacterium seems to be harmed by this relationship, and none of these species can live in similar conditions alone. Some animals eat lichens, including reindeer and some humans—in Northern Europe and Asia, some people still eat a lichen called *Iceland moss*.

(A) What is the best name for the lichen relationship between a fungus and a cyanobacterium?
(B) What ecological succession role do lichens play in areas that have only bare rock?
(C) What is the relationship that a reindeer has with a lichen that it eats?

Natural Cycles and Energy Flow

166. The term *geochemical cycle* describes

(A) the production and breakdown of man-made synthetic materials
(B) the movement of sulfur and other gases from the mantle into the atmosphere
(C) the movement of elements through a repeating series of chemical forms
(D) the movement of magma in the earth's mantle
(E) the transformation of sedimentary rock into igneous rock and back to sedimentary rock

167. The law of conservation of matter holds that

(A) consumers must avoid disposing of petroleum-based materials improperly
(B) when matter is involved in chemical reactions, only a very tiny amount is destroyed
(C) except under very unusual circumstances, matter is never created or destroyed
(D) an object at rest tends to stay at rest, and an object in motion tends to stay in motion
(E) objects fall toward the earth at the same rate, regardless of size

168. If the residence time of N_2 is 400 million years, then

(A) fusion reactions first created the N_2 400 million years ago
(B) each molecule of N_2 remains chemically unchanged for an average of 400 million years
(C) nitrogen's total duration as a nitrate and as a nitrite is an average of 400 million years
(D) no molecule of N_2 lasts longer than 400 million years
(E) N_2 undergoes radioactive decay in approximately 400 million years

169. This element is the basic building block of all organic molecules:

(A) nitrogen
(B) oxygen
(C) phosphorus
(D) hydrogen
(E) carbon

170. Compounds of this element form layers of sedimentary rock on the ocean floor. As one tectonic plate slides under another, much of this sedimentary rock is scraped off, but some can be pulled down into a subduction zone. Heat and pressure change it into a new, gaseous compound, which volcanoes can release into the atmosphere. This element is

(A) carbon
(B) sulfur
(C) nitrogen
(D) phosphorus
(E) calcium

171. Compared to the total amount of carbon in and on the earth now, the amount of carbon present during the planet's formation was

(A) much greater
(B) about the same
(C) much smaller
(D) slightly greater
(E) almost zero

172. Plants directly interact with the biological carbon cycle through

I. respiration
II. photosynthesis
III. decomposition

(A) only I and II
(B) only II and III
(C) only I and III
(D) I, II, and III
(E) only III

173. Sugars and carbohydrates are products of

(A) metabolism
(B) subduction
(C) respiration
(D) decomposition
(E) photosynthesis

174. To construct their shells, phytoplankton get carbon dioxide directly from

(A) the air
(B) the surrounding water
(C) the ocean floor
(D) prey organisms
(E) volcanic vents

175. All of the following are major carbon storage reservoirs EXCEPT

(A) fossil fuels
(B) sedimentary rocks
(C) the air
(D) the earth's core
(E) the oceans

176. Without these organisms, the geological carbon cycle could not take place:

(A) terrestrial photosynthesizing organisms
(B) decomposing organisms and detritus eaters
(C) secondary consumers and top predators
(D) shell-forming marine organisms
(E) lichens and other rock-weathering organisms

177. Unless these two natural phenomena are in balance, global atmospheric carbon dioxide accumulates rapidly:

I. respiration
II. photosynthesis
III. sedimentation
IV. subduction

(A) I and II
(B) II and III
(C) I and IV
(D) III and IV
(E) I and III

178. Deep in the ocean, falling calcium-carbonate shells dissolve at a certain depth, while they collect at a slightly shallower depth. The dividing line between these depths is known as

(A) a thermocline
(B) a lithosphere
(C) the estuarine zone
(D) a lysocline
(E) a trophic level

179. Calcium stored in subsurface rock can return to the surface, but it usually does so as a result of

(A) tectonic plate movement pushing these rock layers upward
(B) erupting volcanoes spewing calcium-rich magma
(C) chemical reactions releasing calcium-rich gas, which diffuses upward
(D) these rock layers slowly dissolving into the oceans
(E) phytoplankton extracting calcium from exposed rock layers

180. Which of the following elements accounts for the greatest proportion of the atmosphere?

(A) nitrogen
(B) oxygen
(C) phosphorus
(D) hydrogen
(E) calcium

181. Nitrogen is an essential element for many biologically important molecules, but organisms can use atmospheric N_2 only once it has been altered by

(A) biomineralization or dissolution in water
(B) volcanic eruptions or reactions with sunlight
(C) the decomposition of organisms or respiration
(D) reactions with acidic gases or photosynthesis
(E) lightning or soil-bacteria processes

182. Which of the following human activities can alter both the carbon and nitrogen cycles at the same time?

(A) allowing artificial fertilizer to run into rivers and streams
(B) growing large numbers of livestock whose waste adds gases to these cycles
(C) constructing dams that obstruct the natural paths of rivers
(D) destroying existing vegetation-rich ecosystems, such as forests
(E) mining large amounts of rock for fertilizer components

183. The term *nitrification* describes which process?

(A) the conversion of N_2 to NO
(B) the conversion of N_2 to NH_3
(C) the conversion of NH_3 to NO_2^- and NO_3^-
(D) the conversion of N_2 to N_2O
(E) the conversion of NO_2^- and NO_3^- into N_2

184. One negative effect of human interference in the nitrogen cycle is
(A) the production of rain containing nitric acid
(B) a decrease in species that thrive on high levels of nitrogen compounds
(C) increased ozone in the stratosphere
(D) a decrease in the total amount of nitrogen present in the atmosphere
(E) an increase in the overall pH of the oceans

185. Carnivores get the majority of their usable nitrogen compounds by
(A) extracting them from the air they breathe
(B) nurturing nitrogen-fixing bacteria in their guts
(C) absorbing them from water
(D) eating herbivores and other carnivores
(E) eating small amounts of soil rich in nitrogen-fixing bacteria

186. Most terrestrial phosphate originates from
(A) freshwater lakes
(B) uplifted rock
(C) phosphate-fixing bacteria
(D) the air
(E) lightning

187. Humans' release of excess phosphates into lakes and rivers can result in
(A) the poisoning of herbivores
(B) phosphate-based gases depleting the ozone layer
(C) acid rain
(D) the overgrowth of aquatic bacteria and algae
(E) global warming

188. Which of the following pollution sources simultaneously releases pollutants containing nitrogen, carbon, and sulfur?
(A) burning fossil fuels
(B) runoff from artificial fertilizers
(C) agricultural animal waste
(D) volcanic eruptions
(E) decaying plastic

189. In which of the following nutrient cycles does the key nutrient remain chemically unchanged throughout?
(A) the sulfur cycle
(B) the nitrogen cycle
(C) the hydrologic cycle
(D) the phosphorus cycle
(E) the carbon cycle

190. The hydrologic cycle is driven by

(A) the earth's rotation
(B) magma convection in the earth's interior
(C) heat from the sun
(D) the moon's gravity
(E) hurricanes and other large atmospheric events

191. Which term describes water moving through soil and permeable rock to groundwater storage aquifers?

(A) infiltration
(B) percolation
(C) condensation
(D) transpiration
(E) transport

192. Of the following hydrologic events, which is most accelerated by warm conditions?

(A) percolation
(B) condensation
(C) transpiration
(D) precipitation
(E) transport

193. Which of the following processes requires condensation nuclei in order to occur?

(A) water vapor condensing into moisture droplets in clouds
(B) liquid water evaporating from land
(C) water droplets becoming precipitation
(D) purification of liquid water by decomposer bacteria
(E) the conversion of precipitation into surface runoff

194. Which of the following terms specifically describes the hydrologic cycle's distribution of water around the world, especially from the oceans to land?

(A) transpiration
(B) precipitation
(C) evaporation
(D) infiltration
(E) transport

195. Approximately what percentage of evaporated water worldwide comes from land?

(A) 50%
(B) 75%
(C) 15%
(D) 40%
(E) 2%

196. Most of the carbon in the earth's crust is stored in

(A) the atmosphere
(B) ocean sediments and sedimentary rocks
(C) fossil fuels
(D) plants
(E) the soil

197. Which of the following lists begins with the largest calcium reservoir and ends with the smallest?

(A) atmosphere, oceans, soil
(B) soil, atmosphere, oceans
(C) oceans, soil, atmosphere
(D) oceans, atmosphere, soil
(E) soil, oceans, atmosphere

198. Nitrification results in the two final products

(A) ammonia and nitrogen gas
(B) nitrate ions and nitrite ions
(C) DNA and proteins
(D) ozone gas and oxygen gas
(E) nitrous oxide and nitric oxide

199. Of the following human activities, which two add sulfur dioxide to the environment?

I. burning fossil fuels
II. smelting ores to obtain metals
III. disposing of radioactive waste
IV. releasing fertilizer into rivers

(A) I and II
(B) II and III
(C) III and IV
(D) II and IV
(E) I and IV

200. Which of the following trophic levels contain animals that eat herbivores?

I. tertiary consumers
II. secondary consumers
III. primary consumers
IV. producers

(A) I and II
(B) II and III
(C) III and IV
(D) I and III
(E) II and IV

201. Volcanoes release a great deal of gas when they erupt, but the composition of those gases depends on the magma type. A certain volcano in the Pacific Ocean releases about 49% carbon dioxide, 37% water vapor, 12% sulfur dioxide, and 1.5% carbon monoxide.

(A) How does carbon cross over from the biological carbon cycle to this part of the geological carbon cycle?
(B) What is the process by which sulfur dioxide from this volcano can become a component of organisms?
(C) How does carbon make its way from volcanic carbon dioxide emissions into organisms?

202. After about a century of logging, deforestation in Madagascar's rain forests and its mangroves on the coast have left the country with terrible erosion problems. Slash-and-burn agricultural practices have turned a once verdant island into one with some of the fastest-changing coastline—sediment washes down Madagascar's large rivers and fills coastal waterways on its way to the ocean. According to NASA, astronauts say it looks like the country is "bleeding" after heavy rain washes its red soil out to sea.

(A) How is the nitrogen cycle on Madagascar likely affected by its deforestation?
(B) How is the hydrologic cycle on Madagascar likely affected?
(C) How is the biological carbon cycle on Madagascar likely affected?

203. In the limestone caves of Illinois, food comes from only a few sources. Heavy rains wash plants and other organic debris into the caves occasionally, but most of the food comes from outside by way of animals. Sometimes animals die deep in these caves, but much of the underground ecosystem depends on their droppings. Raccoon, frogs, cave crickets, and especially bats forage outside for food and leave fecal matter on cave floors. Bacteria and fungi break down these droppings into their mineral constituents and are themselves eaten by animals, such as millipedes and flatworms. Spiders, salamanders, and other animals catch and eat these animals when possible, but most cave dwellers can survive for a very long time without any food at all.

(A) What producers does this cave ecosystem depend upon?
(B) Which of these species are the cave equivalent of primary consumers?
(C) Which of these species are the secondary consumers?

Population Biology and Dynamics

204. Which of the following lists of terms is organized from most specific to most inclusive?

(A) community, population, ecosystem
(B) population, community, ecosystem
(C) species, ecosystem, population
(D) individual, community, population
(E) individual, species, population

205. If a population has high fecundity but low fertility,

(A) the current number of offspring is relatively low and is likely to stay low
(B) the current number of offspring is relatively high, but it is likely to decrease or stay the same
(C) the relative number of offspring is likely to increase and decrease wildly with time
(D) the current number of offspring is relatively low, but it may greatly increase
(E) the current number of offspring is relatively high and is likely to stay high

206. This term describes the number of years an individual is expected to live at the time it is born:

(A) life span
(B) fecundity
(C) fertility
(D) life expectancy
(E) mortality

207. When the individuals of a population are most likely to be found in groups, their physical distribution is known as

(A) uniform dispersion
(B) exponential growth
(C) random dispersion
(D) clumping
(E) genetic drift

208. The following factors contribute to an increase in a population's size:

I. births
II. immigration
III. emigration

(A) I only
(B) II only
(C) I and II
(D) II and III
(E) I and III

209. When a population exceeds the environment's carrying capacity (*K*),

(A) it can suffer a sudden population crash
(B) it enters the exponential growth phase
(C) it overcomes environmental resistance
(D) its age structure becomes heavily weighted in favor of older individuals
(E) it changes its distribution from uniform dispersion to random dispersion

210. When a population's growth adjusts to approximate the environment's carrying capacity (*K*), it is known as

(A) a species' biotic potential
(B) exponential growth
(C) a population crash
(D) logistic growth
(E) irregular behavior of population growth

Questions 211–215 refer to the two curves plotted on the graph in Figure 7.1. They represent two different populations undergoing different patterns of change over time. Lettered points on each curve represent different moments in each group's population changes.

211. At point V, Population 1 is

(A) slowing its growth
(B) overshooting the environment's carrying capacity
(C) growing exponentially
(D) experiencing environmental resistance
(E) growing logistically

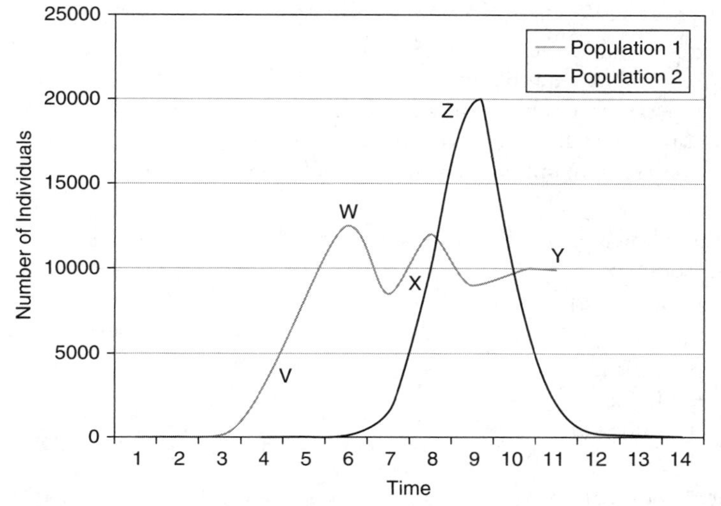

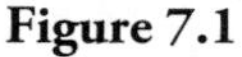
Figure 7.1

212. What is the approximate carrying capacity of the environment experienced by Population 1?

(A) 8,000
(B) 10,000
(C) 13,000
(D) 15,000
(E) 20,000

213. Between points W and Y, Population 1 is experiencing

(A) a population crash
(B) exponential growth
(C) geometric growth
(D) its intrinsic rate of increase
(E) logistic growth

214. At point Z, Population 2 is

(A) overshooting the environment's carrying capacity
(B) growing exponentially
(C) undergoing a population crash
(D) experiencing logistic growth
(E) experiencing no resource limitations

215. After point Y, Population 1 will most likely

(A) find a stable size at about 5,000
(B) increase exponentially
(C) experience a population crash
(D) continue to fluctuate near its current level
(E) eventually maintain a constant population at 10,000

216. Of the following, which is an example of a density-independent population control?

(A) limited food
(B) limited space
(C) parasitism
(D) earthquakes
(E) infectious disease

217. When a population size tends to fluctuate around the environment's carrying capacity, biologists consider it to be

(A) irruptive
(B) stable
(C) irregular
(D) decaying
(E) crashing

218. As demographers define it, industrial population growth includes

(A) high birthrates and high death rates
(B) high birthrates and low death rates
(C) low birthrates and high death rates
(D) low birthrates and low death rates
(E) birthrates and death rates that increase and decrease unpredictably

219. As opposed to a K-selected species, an r-selected species usually has

(A) numerous relatively small offspring
(B) offspring that tend to survive to reproductive age
(C) a population that stays close to the environment's carrying capacity
(D) high parental care of offspring
(E) a relatively low population growth rate

220. A population of island birds has a very homogeneous genetic pool. Which of the following can result in greater genetic diversity?

(A) a small number of birds leaving to start a new colony
(B) a bottleneck event
(C) genetic drift
(D) inbreeding
(E) a sudden die-off among one generation of the birds' offspring

221. Humans can increase an area's biodiversity when they

(A) drain a swamp to build a tree farm
(B) plow a field to plant crops
(C) break one large habitat into two smaller ones
(D) encourage a river's algae growth by allowing an inflow of fertilizer
(E) allow two populations of a species to trade members

222. A new island is colonized by a small number of mice that occasionally trade members with mainland mice when the receding tide reveals a land bridge between the two. This is an example of

(A) a bottleneck event
(B) a metapopulation
(C) the founder effect
(D) genetic drift
(E) inbreeding

223. Demographers expect most developed nations to face challenges by the middle of this century involving which of the following?

I. high birthrate
II. rising death rate
III. high emigration
IV. negative growth

(A) I and II
(B) I and III
(C) II and III
(D) II and IV
(E) III and IV

224. Of the following characteristics, which three are typical of K-selected species?

I. late successional colonization
II. early reproductive age
III. generalist niche
IV. small size in adulthood
V. few but large offspring
VI. specialist niche

(A) I, II, and III
(B) II, III, and IV
(C) II, IV, and V
(D) I, V, and VI
(E) II, IV, and VI

225. Depending partly on their reproductive strategies, different organisms have different life expectancies described by survivorship curves. The three general survivorship curves are

I. constant loss
II. erratic loss
III. early loss
IV. logarithmic loss
V. late loss

(A) I, II, and V
(B) II, III, and IV
(C) I, III, and V
(D) II, III, and V
(E) I, II, and III

Questions 226–231 refer to the two age-structure graphs in Figure 7.2. Each represents the number of people of different age-groups living in a country at the same time.

226. Using only these graphs, it's possible to deduce that compared to Country Y, Country X currently has a higher

(A) emigration rate
(B) immigration rate
(C) birthrate
(D) intrinsic rate of increase
(E) population density

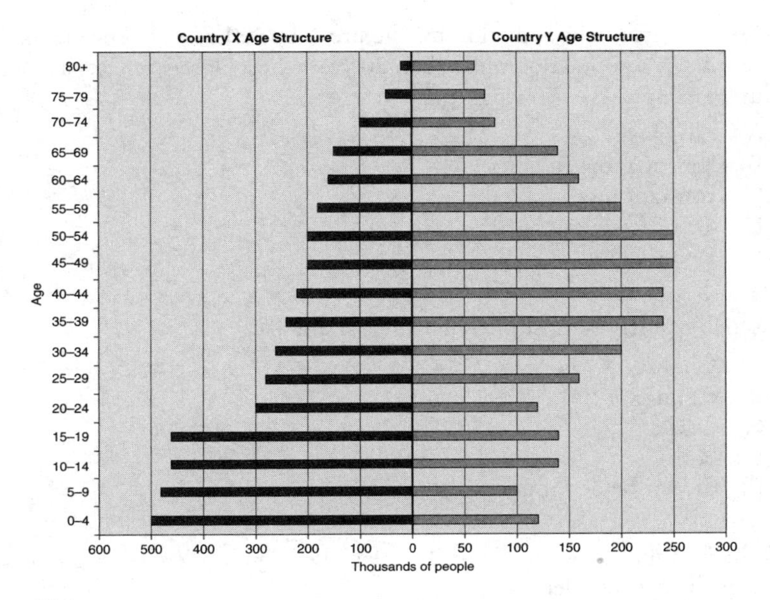

Figure 7.2

227. If Country Y is a typical developed nation, in 10 to 20 years it will begin suffering from

(A) a reduced tax base
(B) a high birthrate
(C) high population density
(D) low emigration
(E) a founder effect

228. If Country X has recently industrialized, in 15 years its growth rate will most likely

(A) decrease
(B) increase
(C) remain the same
(D) fluctuate
(E) decrease only slightly

229. Assuming its total population for the previous 200 years has been constant, Country X's population older than 20 years most closely reflects a survivorship curve showing

(A) early loss
(B) logistic growth
(C) constant loss
(D) logarithmic growth
(E) a population crash

230. Which country is most likely postindustrial?

(A) Country X
(B) Country Y
(C) neither
(D) both
(E) cannot be determined with these diagrams

231. Which of the following factors best explains Country Y's lower population in age brackets under 30 years old?

(A) emigration
(B) immigration
(C) a lower birthrate
(D) a higher death rate
(E) an increased life expectancy

232. In contrast to humans, animals' fecundity is often very close to their

(A) life expectancy
(B) fertility
(C) life span
(D) survivorship
(E) mortality

233. One hundred mockingbirds live on a five-square-kilometer island, and in one year, they produce 20 offspring. In the same time, 10 birds die, and 30 birds immigrate. What is the island's new mockingbird population density?

(A) 150 individuals per square kilometer
(B) 140 individuals per square kilometer
(C) 28 individuals per square kilometer
(D) 16 individuals per square kilometer
(E) 8 individuals per square kilometer

234. Starting in the 1990s, the U.S. wild pig population exploded. Escaped domestic pigs and wild boars interbred, and their populations grew exponentially, so that by 2008, the U.S. Department of Agriculture estimated that nearly four million of the animals ran wild in more than 30 states, with the largest populations in Texas, Florida, and California. They are highly adaptable to many habitats and have a high reproductive rate, with females becoming fertile at about eight months of age and producing two litters every year, each with about six small piglets. Although sows carefully protect their brood, juveniles have a very high mortality rate, while adults have a lower mortality rate after one year. Adult wild pigs grow to more than 45 kg—about 100 pounds—and live an average of five years. True omnivores, wild pigs eat fruit, seeds, nuts, grubs, small animals, and occasionally smaller pigs. They destroy crops and are dangerous to humans, so states including Missouri have eased regulations to encourage hunters to shoot wild pigs. Other carnivores of wild pigs include wolves, bears, coyotes, birds of prey, bobcats, and cougars.

(A) Do wild pigs conform more to the characteristics of r-selected or K-selected species? In what ways do they differ from the standard for that type of species?
(B) What sort of survivorship curve best describes the mortality of wild pigs throughout their lives?
(C) During the 1990s, had wild pigs approached their environment's carrying capacity?
(D) What elements in the paragraph in Question 234 comprise the environmental resistance that helps to establish wild pig carrying capacity?

235. According to the U.S. Census Bureau, between 2004 and 2005, the U.S. Northeast had 362,000 immigrants from elsewhere in the United States, while 774,000 people emigrated. An additional 323,000 people immigrated into the Northeast from other countries. During the same period, the U.S. West had 718,000 immigrants from inside the country and 655,000 people emigrating, and 597,000 people immigrated to the West from other countries. In 2004, there were about 148.9 million women in the United States and a total of 293 million people. About 61.6 million U.S. women were of reproductive age in 2004, as defined by the Census Bureau—that is, they were 15 to 44 years old. About 3.7 million of these women had given birth in the previous year.

(A) What is the net 2003–2004 population change in the U.S. Northeast from all immigration and emigration?
(B) Considering the U.S. Northeast and West, which region had greater overall immigration? Which had greater overall emigration?
(C) What was the fertility of U.S. women between 2003 and 2004?
(D) What was the fecundity of U.S. women between 2003 and 2004?

236. In 1665, the Great Plague killed as many as 80,000 people in London, England, as plague-carrying fleas spread the disease among the city's inhabitants. The next year saw the Great Fire of London, which destroyed whole regions of the city and killed a reported six people—but probably a great many more. Some historians claim that the fire stopped plague-ridden fleas by killing their host rats by the thousands. Many historians believe that a "Little Ice Age" affected Northern Europe and other parts of the world, decreasing crop harvests and increasing instances of starvation and famine from about 1550 until about 1850.

(A) Was the Great Plague a density-independent or density-dependent factor affecting the human population of England?

(B) Which of the events mentioned above represent a density-independent factor affecting the human population?

(C) Which of the events mentioned above are abiotic factors, and which are biotic?

Agriculture and Aquaculture

237. Vitamin A deficiency can cause

(A) anencephaly
(B) anemia
(C) blindness
(D) goiter
(E) obesity

238. An insufficient amount of protein and calories in a person's diet can result in

(A) marasmus
(B) kwashiorkor
(C) drowsiness and inactivity
(D) abdomen swollen with fluid
(E) gout

239. Obesity is

(A) when someone is more than 20 pounds over ideal weight
(B) when someone is more than 30 pounds over ideal weight
(C) occurring in about one-quarter of the U.S. population
(D) the result of consuming too many whole grains and vegetables
(E) occurring in about 62% of the population today

240. Which of these crops are mainly GMO crops?

(A) soybeans, cotton, and corn
(B) wheat, rice, and corn
(C) wheat, cotton, and corn
(D) potatoes, wheat, and rice
(E) soybeans, rice, and cotton

241. The ideal method of irrigation for water conservation is

(A) flood irrigation
(B) ditch irrigation
(C) sprinkler irrigation
(D) drip irrigation
(E) canal irrigation

242. Bt, or *Bacillus thuringiensis,* belongs to which category of pesticide?

(A) inorganic
(B) natural organic
(C) organophosphates
(D) microbial/biological agents
(E) carbamates

243. Agriculture that alternates two or more crops to reduce wind and water erosion is known as

(A) terracing
(B) shelterbelts
(C) soil compaction
(D) contour farming
(E) strip cropping

244. The types of crops that are used directly by a farmer or sold locally are

(A) cash crops
(B) ruminant crops
(C) subsistence crops
(D) monoculture crops
(E) swidden crops

245. Which of the following types of crops have the benefit of conserving soil, retaining nutrients, and saving energy?

(A) annual crops
(B) cash crops
(C) subsistence crops
(D) monoculture crops
(E) perennial crops

246. Which continent has the greatest percentage area of agriculture?

(A) Asia
(B) Europe
(C) North America
(D) Australia
(E) Africa

247. The National Plant Germplasm System provides

(A) antimicrobial agents
(B) cold storage for seeds
(C) screening for GMOs
(D) payments to farmers
(E) protection of soil fertility

248. Which of the following agencies is responsible for destroying shipments of food that exceed pesticide limits?

(A) FDA
(B) USDA
(C) EPA
(D) A and B
(E) DDT

249. Water pollution and breeding between farmed and wild types put which species on the endangered list?

(A) Atlantic salmon
(B) Pacific salmon
(C) tuna
(D) Atlantic cod
(E) Peruvian anchovy

250. The optimal growth of a crop can be prevented by a shortage in the soil of a chemical element known as

(A) a macronutrient
(B) a micronutrient
(C) a limiting factor
(D) a fumigant
(E) Liebig's law

251. In China, desertification has been offset by

(A) logging the virgin forest
(B) aquaculture of tuna
(C) building dams
(D) diverting rivers
(E) planting trees

252. Topsoil, the layer of soil that supports crops, is also known as the
(A) litter layer
(B) A horizon
(C) zone of accumulation
(D) C horizon
(E) soil profile

253. An agricultural field that is not harvested for at least a season is
(A) subsidized
(B) terraced
(C) contoured
(D) fallow
(E) tilled

254. Elevated levels of carbon dioxide from global warming have which of the following effects?
(A) insects eating fewer plants
(B) decreased nitrogen in plants
(C) increase in bird populations
(D) increased nitrogen in plants
(E) increase in available farmland

255. Approximately what percentage of crops are destroyed by pests, spoilage, and disease?
(A) 30%
(B) 10%
(C) 50%
(D) 20%
(E) 5%

256. Herbicides account for what percentage of pesticides in U.S. waters?
(A) 80%
(B) 10%
(C) 60%
(D) 5%
(E) 20%

257. *Silent Spring* is a book published in 1962 that warned of the impact of
(A) deforestation in the Amazon
(B) pesticides and herbicides
(C) nuclear war
(D) population growth
(E) drought on agriculture

258. The Clean Water Act was enacted in

(A) 1977
(B) 1970
(C) 1963
(D) 1912
(E) 1989

259. How many years ago did agriculture and the domestication of animals begin?

(A) 5,000
(B) 13,000
(C) 250
(D) 500,000
(E) 10,000

260. Which of the following is NOT a way that agriculture affects climate?

(A) changing of land cover
(B) increase of carbon dioxide
(C) use of fire to clear land
(D) increase of biological diversity
(E) production/use of fertilizers

261. The diagram in Figure 8.1 illustrates

(A) primary succession
(B) secondary succession
(C) tertiary succession
(D) pioneer plants
(E) no-till agriculture

Figure 8.1

262. Which of the following is true about people in China?

I. They are consuming more meat products than in the past.
II. More grain production is required.
III. China has more agricultural land available now than in the past.

(A) I only
(B) II only
(C) I and II only
(D) II and III only
(E) I, II, and III

263. Minimum tillage is a farming technique that

(A) decreases the use of herbicides
(B) increases soil compaction
(C) reduces soil erosion
(D) increases evaporation of water
(E) increases energy consumption

264. An organic fertilizer that is made from plowing under plants instead of harvesting them is known as

(A) green manure
(B) cow manure
(C) synthetic fertilizer
(D) leguminous plants
(E) green herbicides

265. Pasture is land that

(A) is burned to provide room for agriculture
(B) is left fallow for a season
(C) is not used for ruminants
(D) is planted with forage for animals
(E) provides food for animals without planting

266. Which of the following requires the most water to produce?

(A) potatoes
(B) rice
(C) corn
(D) chicken
(E) beef

267. Desertification is occurring in many parts of the world due to the impact of people and climate change.

(A) Describe three signs of desertification in an area.
(B) Describe two specific examples of how people have contributed to desertification.
(C) Describe alternative methods that could be used instead of the ones that were described in (B).

268. Overfishing is quickly decreasing the world's supply of fish, which would greatly impact the availability of protein to developing nations.

(A) Describe two fishing methods that contribute to the problem of overfishing.
(B) Describe two methods that would help prevent overfishing.

269. Farmer Jill started growing organic vegetables in her backyard and was so successful that she decided to purchase land and fulfill her dream of owning her own organic farm. While looking at properties, she found three properties for sale. The first is currently run as a large industrial farm but is not organic. The second is a small farm that used resource-based methods of agriculture. The third piece of land is not yet farmed.

(A) Assume she chooses the first property. Describe two problems she would have with changing this land to an organic farm.
(B) Describe what features she would see at the second property that directly relate to using the resource-based methods.
(C) Assume she purchases the third property. Identify and describe
  (i) organic methods or concepts that would be used in designing and building the new organic farm
  (ii) the drawbacks of owning an organic farm
  (iii) the principles and methods of integrated pest management that would be used on the organic farm

Forestry and Rangelands

270. The major determinants of forest type are

(A) tree species and animal species
(B) temperature and rainfall
(C) elevation and proximity to water
(D) tree species and elevation
(E) tree species and rainfall

271. A temperate rain forest, such as the one on the west coast of North America, primarily has

(A) conifer trees, wet climate, cool temperatures
(B) flat-leaf trees, wet climate, warm temperatures
(C) conifer trees, dry climate, cool temperatures
(D) flat-leaf trees, wet climate, cool temperatures
(E) conifer trees, wet climate, warm temperatures

272. What percentage of the earth's original rain forests have been cleared?

(A) 25%
(B) 90%
(C) 10%
(D) 50%
(E) 32%

273. Which method of timber harvesting cuts a small percentage of mature trees every 10 to 20 years?

(A) clear-cutting
(B) deforestation
(C) selective cutting
(D) old-growth harvesting
(E) industrial logging

274. The USDA cites the greatest number of threatened plant species in which biome?

(A) tropical rain forest
(B) desert
(C) arctic tundra
(D) temperate forest
(E) grassland

275. Semiarid grasslands, with a mean rainfall of between ¼ and ½ cm per year, are also known as

(A) steppes
(B) taigas
(C) deserts
(D) forests
(E) chaparrals

276. Which of the following are benefits of forest fires?

I. pinecones open and release seeds
II. removal of brush and replenishment of soils
III. increasing the number of crown fires

(A) I only
(B) II only
(C) I and II only
(D) II and III only
(E) I, II, and III

277. The difference between drought and degradation of land is

(A) degradation is not affected by climate
(B) degradation results from unsound human activities
(C) drought is an unnatural event
(D) degradation increases economic yield
(E) drought increases biodiversity and economic yield

278. The landscape perspective of environmental science involves

(A) the protection of local areas such as a stand of trees
(B) concern for the appearance of the environment
(C) preserving only what is immediately visible
(D) protecting individual species by protecting their ecosystems
(E) protecting land from agricultural use

279. The use of nitrogen by a plant allows it to make

(A) proteins
(B) sugar
(C) CO_2
(D) O_2
(E) water

280. The Public Rangelands Improvement Act of 1978 was

(A) written by ranchers to increase area for grazing
(B) passed by Congress to reduce grazing where there was damage
(C) written to improve Native American lands
(D) responsible for the deterioration of forests
(E) written to prohibit building houses on grazing land

281. When a forest is harvested at a rate that does not decrease supply year after year, it is said to be

(A) clear-cut
(B) sustainable
(C) old-growth
(D) suppressed
(E) deforested

282. In what year was the U.S. Forest Service established?

(A) 1968
(B) 1925
(C) 1905
(D) 1918
(E) 1936

283. What is "an area where the earth and its community of life are untrammeled by man, where man himself is a visitor who does not remain" as defined by U.S. law?

(A) biome
(B) rangeland
(C) commons
(D) hectare
(E) wilderness

284. The purpose of the Wilderness Act of 1964 was

(A) to change to different methods of timber harvesting in forests
(B) to set aside land owned by the Bureau of Land Management
(C) to protect tropical rain forests in Guam
(D) to set aside primitive forest land for protection
(E) to support businesses promoting such activities as skiing and snowmobiling in wilderness areas

285. Indirect deforestation is

(A) death of trees from pollution or disease
(B) killing some trees unintentionally during selective cutting
(C) loss of trees from logging
(D) cutting of forests by local people for fuel
(E) timber harvesting of plantation forests

286. The bottom half of a grass plant that provides food for the plant's roots is known as the

(A) photosynthetic process
(B) rangeland
(C) primary forest
(D) clear-cut area
(E) metabolic reserve

287. The trees that grow in the understory of the forest are known as

(A) intermediate
(B) dominants
(C) suppressed
(D) second-growth forest
(E) codominants

288. Which of the following is NOT a public service function of forests?

(A) slowing of erosion
(B) increasing evaporation of water
(C) providing recreational activities
(D) providing timber
(E) providing habitat for wildlife

289. What percentage of tropical forest logging is done in a sustainable manner?

(A) 0.1%
(B) 1%
(C) 10%
(D) 25%
(E) 50%

290. Trees are adapted to their own particular environmental conditions, also known as a(n)

(A) old-growth forest
(B) landscape
(C) niche
(D) community forest
(E) plantation

291. What makes grasslands so suitable for agriculture?

(A) They are flat and easier to plow.
(B) High-quality soil results from a deep organic layer.
(C) There is a large area available in the world.
(D) There are no animals that eat the grasses.
(E) There are few fires.

292. When were domestic animals such as cows brought to North America?

(A) 1800s
(B) 1900s
(C) 1400s
(D) 1200s
(E) 1600s

293. Feedlots are damaging to the environment because of

(A) overgrazing
(B) deforestation
(C) erosion
(D) pollution
(E) desertification

294. The majority of wood in forests of developing countries is used for

(A) timber for construction
(B) firewood
(C) paper
(D) furniture
(E) pulp

295. The primary purpose of parks is for

(A) conservation
(B) research
(C) people
(D) hunting
(E) isolation

296. The habitat for spotted owls is

(A) old-growth forests
(B) taiga forests
(C) rangelands
(D) rain forests
(E) second-growth forests

297. The maximum number of trees that a piece of land can produce within a certain time frame describes its

(A) maximum harvest
(B) silviculture quota
(C) certification of forestry
(D) lumber limit
(E) site quality

298. Brief, intense grazing done by wild herds in which a herd is confined to a small area, eats everything, and fertilizes heavily before moving on is known as

(A) selective cutting
(B) clear-cutting
(C) overgrazing
(D) rotational grazing
(E) desertification

299. Forests are disappearing from the world at a fast rate that can be remedied only by better forest management.

(A) Describe the four different methods of tree harvesting: clear-cutting, strip cutting, selective cutting, and shelterwood cutting.
  (i) Explain the process of each method of harvesting.
  (ii) Describe the environmental drawbacks of each.
  (iii) Describe the benefits of each.
(B) Describe two methods of maintaining sustainable forestry.
(C) Discuss two pieces of legislation that were passed to protect forests in the United States.

300. Forest fires are often viewed in a negative light—they threaten lives, burn down beautiful stands of trees, and endanger towns. However, forest fires have many benefits as well that are often overlooked.

(A) Describe three benefits that forest fires can provide.
(B) Discuss the purpose of prescribed fires.
(C) Assume that you own a home in an area that is prone to forest fires. Explain the steps you would take to reduce forest fires on your property.

301. Rangeland covers approximately 70% of the earth's surface, yet most of it is in poor condition. Grazing practices are responsible for much of the degradation.

(A) Compare and contrast traditional herding of cattle with industrial methods of raising cattle.

(B) Describe two effects of introducing nonnative grazing animals to an area of rangeland.

(C) You have inherited the family cattle business from your grandfather in Texas. You want to run the business in a sustainable way. Describe five methods you can use to make the business more ecologically sound.

Land Use

302. People can be "pulled" to emigrate from rural areas into urban areas by

(A) poverty in rural areas
(B) lack of agricultural land outside of cities
(C) lack of jobs outside of cities
(D) available housing in urban areas
(E) war in the countryside

303. If 20,000 of a small country's total population of 1 million live in cities, the country's degree of urbanization is

(A) 20%
(B) 5%
(C) 2%
(D) 0.2%
(E) 0.5%

304. If a country maintains the same total population of 100 but its rural population shrinks from 55% to 52% in one year, what is the urban growth rate for this period, assuming that the rest of the population is composed of city dwellers?

(A) 2.7%
(B) 6.7%
(C) 3.0%
(D) 0.9%
(E) 6.3%

305. Of the following statements, which one accurately reflects current urban growth patterns worldwide?

(A) Urban growth is slower in developed countries than in developing countries.
(B) Urban poverty is becoming less prevalent.
(C) City dwellers are staying at a constant percentage of the world population.
(D) In developing countries, the rural population is becoming a greater proportion of the population.
(E) The number of megacities worldwide is shrinking slightly.

306. Which of the following statements accurately describes urban growth patterns in the United States?

(A) Between 1850 and 1900, urban growth stagnated.
(B) During the mid-20th century, suburbs and small cities lost more people to larger nearby cities than they gained.
(C) Ongoing migration inside the country is generally headed toward the Southwest.
(D) The majority of the rural population has migrated to urban areas from rural areas since the 1970s.
(E) Since the 1980s, the birthrate has been largely responsible for the country's growing population.

307. All of the following statements are examples of the negative effects of urban sprawl EXCEPT

(A) an increase in obesity in sprawl areas
(B) the loss of central city tax bases
(C) the loss of prime cropland, grass, and forests
(D) decreased water runoff
(E) increased surface-water pollution

308. In the United States, urban sprawl was encouraged by the federal government through

I. loan guarantees for veterans buying new single-family homes
II. federal funding of highways
III. the introduction of home-loan deductions for federal income tax
IV. large federally funded suburban developments outside western U.S. cities

(A) I, II, and III
(B) II, III, and IV
(C) I, III, and IV
(D) only I and II
(E) only II and IV

309. People in urban areas often have advantages over rural dwellers, including

(A) lower chances of exposure to hazardous wastes
(B) lower levels of asthma and other respiratory problems
(C) better access to health care and other services
(D) exposure to fewer sources of noise pollution
(E) living where infectious diseases are less able to propagate

310. Cities can put extra stress on the environment's water resources by

I. depriving nearby, less-developed areas of surface water and groundwater
II. decreasing the amount of water entering soil and groundwater by increasing runoff
III. decreasing nearby cloud formation by producing excess heat and CO_2
IV. contributing to water pollution

(A) I, II, and III
(B) II, III, and IV
(C) I, II, and IV
(D) only I and III
(E) only III and IV

311. In the United States, rail transportation compares favorably to automobile transportation in all of the following areas EXCEPT

(A) flexible transportation schedules
(B) air pollution
(C) energy efficiency
(D) accidental injuries and deaths
(E) contributing to traffic congestion

312. Some U.S. cities including Portland, Oregon, apply zoning, planning, tax breaks, building regulations, and other policies to control sprawl and protect the local environment. This approach is called

(A) fundamental land
(B) reconciliation ecology
(C) cluster development
(D) smart growth
(E) land-use planning

313. One approach to building more efficient housing with recreational space involves concentrating high-density housing in one part of a site and leaving a large proportion for shared open space. This is known as

(A) fundamental land
(B) land-use planning
(C) smart growth
(D) mixed-use development
(E) cluster development

314. Urban planning experts call a neighborhood with stores, light industries, offices, high-density housing, and mass transportation within walking distance

(A) a mixed-use development
(B) smart-growth planning
(C) cluster development
(D) reconciliation ecology
(E) land-use planning

315. In China, developers can face the death penalty for illegally building on arable areas that are designated as

(A) mixed-use zoning
(B) new urban areas
(C) cluster developments
(D) fundamental land
(E) greenbelt

316. Transit corridors connect cities such as Toronto to smaller nearby urbanized areas, with the rest of the area surrounding the city center devoted to recreational open space known as

(A) fundamental land
(B) mixed-use zoning
(C) greenbelt
(D) cluster development
(E) reconciliation ecology

317. The amount of biologically productive area required to support a person and dispose of his or her waste is known as

(A) the free-access resource area
(B) the common-property resource area
(C) the ecological capacity
(D) a sustainable yield
(E) an ecological footprint

318. An equation to estimate a human population's ecological impact uses the variables *I* (total impact), *P* (population size), *A* (population affluence), and *T* (population technology level). Using that equation, if *P* and *A* both double compared to a previous situation but *T* remains the same, what happens to *I*?

(A) *I* drops to zero.
(B) *I* drops to half its previous value.
(C) *I* remains the same value.
(D) *I* increases to four times its previous value.
(E) *I* cannot be calculated in this case.

319. Flooding is often a bigger problem in cities than in rural areas due to which of the following factors?

I. Cities are often built in coastal areas or on floodplains.
II. Cities' greater proportion of impervious cover increases runoff.
III. City-related development often damages wetlands that can absorb excess water.
IV. Cities release excess heat and carbon dioxide, which increase local rainfall.

(A) only I and II
(B) only II and III
(C) only I and IV
(D) I, II, and III
(E) II, III, and IV

320. Sewage runoff from human settlements into the oceans can cause all of the following EXCEPT

(A) harmful algal blooms
(B) exposure of marine life and people to high levels of infectious bacteria
(C) unusually high levels of human viruses in the water
(D) oxygen-depleted water zones that suffocate marine life
(E) calcium-rich zones that damage fish and shellfish by hardening and stiffening joints

321. Measures that can decrease the prevalence of oxygen-depleted zones include all of the following EXCEPT

(A) reducing the emission of nitrogen compounds by motor vehicles
(B) constructing industrial-scale saltwater oxygenating plants
(C) improving sewage treatment to reduce the emission of nitrogen compounds into the water
(D) reestablishing coastal wetlands to help absorb excess nitrogen compounds
(E) planting vegetation to absorb extra nitrogen before it reaches rivers, streams, and oceans

322. Which of the following lists of land uses is arranged in order from the largest share of the earth's terrestrial area to the smallest?

(A) urban space, agricultural land, wooded areas
(B) agricultural land, wooded areas, urban space
(C) wooded areas, agricultural land, urban space
(D) urban space, wooded areas, agricultural land
(E) wooded areas, urban space, agricultural land

323. Most U.S. federal public land is located in the state of

(A) New York
(B) Utah
(C) California
(D) Alaska
(E) Colorado

324. In the United States, parcels of land managed by the Forest Service and the Bureau of Land Management are used for many of the same purposes, with important differences. Unlike land managed by the Forest Service, land under the Bureau of Land Management is NOT often used for

(A) mining
(B) natural gas extraction
(C) livestock grazing
(D) oil extraction
(E) recreation

325. The U.S. Fish and Wildlife Service oversees 542 tracts of land intended mostly for

(A) protecting habitats for game birds and mammals
(B) oil and gas extraction
(C) recreation
(D) historical site preservation
(E) watershed conservation

326. Each set of federally managed areas in the United States has different restrictions on its use. The U.S. federal lands with the most restrictions are known as the

(A) National Forest System
(B) National Resource Lands
(C) National Wildlife Refuges
(D) National Park System
(E) National Wilderness Preservation System

327. Which of the following public land conservation principles have developers and resource-extraction industries in the United States actively opposed?

I. Protecting the biodiversity should be the primary goal.
II. Resource extraction should not be government subsidized.
III. Resource extractors should pay fair compensation for resources removed.
IV. Everyone using public lands should be liable for environmental damage.

(A) only I
(B) I, II, and III
(C) I, II, III, and IV
(D) II, III, and IV
(E) I, III, and IV

328. Biologist Garrett Hardin reasoned that each person degrading free-access, renewable resources feels that an individual person's impact is nearly imperceptible, that another person is bound to use the same resource in any case, and that as a renewable asset, the resource will rebound. Hardin called the large-scale effect of this mind-set

(A) the tragedy of the commons
(B) distorted risk perception
(C) resistance-to-change environmental management
(D) the reversibility principle
(E) short-sighted stewardship

329. Approximately 3.8 billion hectares of earth's original forests remain. This number represents what proportion of the original total?

(A) about 5%
(B) about 20%
(C) about 50%
(D) about 70%
(E) about 90%

330. "A natural, inorganic solid having a particular crystalline structure and chemical composition" is the definition of

(A) an element
(B) a mineral
(C) a compound
(D) a gemstone
(E) a fossil fuel

331. The process of using water to wash soil and other unwanted materials to reveal desired minerals is known as

(A) strip mining
(B) smelting
(C) heap-leach extraction
(D) placer mining
(E) water mining

332. In 2011, the Australian government produced a report on urbanization suggesting that development in the country should concentrate on filling in urban areas, particularly in derelict sites. Inner cities, said the report, should be connected to transport corridors and green spaces to provide a high standard of living in a more efficient manner. At the same time, Prime Minister Julia Gillard promised to increase job opportunities in outer suburbs, which have fewer community services, such as health, education, and housing facilities.

(A) In the paragraph above, which stated goal is likely to increase urban sprawl?
(B) How might filling in urban areas allow people to experience a high standard of living more efficiently?
(C) How might urban areas with more community services better survive economic fluctuations?

333. In 2009, the PacRim mining company proposed its Chuitna Coal Project in Alaska—a strip mine and associated development that together would cover about 78 square kilometers (about 30 square miles) of the Beluga Coal Fields near the Cook Inlet southwest of Anchorage. Many activists, residents, and fishery experts are concerned that although the company is taking some steps to reduce waste released, the mine will destroy area streams, which are important salmon hatcheries that contribute to the local economy. Mine proponents estimate that the 25-year first phase of the project could create as many as 350 new jobs and stimulate the local economy by involving local businesses. The company also plans to create a major port nearby that could encourage further development of the area, as well as further mining.

(A) What is strip mining?
(B) How might strip mining endanger streams and the salmon inhabiting them?
(C) At what point would the Surface Mining Control and Reclamation Act of 1977 require PacRim to attempt some environmental restoration?

334. Officially created in 1872, Yellowstone National Park was the first of 58 national parks in the United States and the first national park in the world. Its 8,987 square kilometers (or 3,472 square miles) contain breathtaking beauty, natural wonders, and habitats featuring grizzly bears, bison, and rare thermophilic bacteria. About 80% of Yellowstone is covered by forest, with 15% devoted to grassland and 5% devoted to water.

(A) Why did the United States first establish national parks such as Yellowstone?
(B) How do parks help to preserve endangered and rare species?
(C) What U.S. agency administers the country's national parks?
(D) What are some of the chief threats to national parks?

Energy Consumption

335. Compared with the "Calorie" unit that gauges food energy, how large is the "calorie" unit that scientists use for other types of energy?

(A) 1/1,000 times as large
(B) 1/10 as large
(C) the same size
(D) 10 times larger
(E) 100 times larger

336. Carrying an apple across the room and bringing a pot of water to boil are both best described as examples of

(A) energy
(B) potential energy
(C) electromagnetism
(D) work
(E) material efficiency

337. Which of the following is best described as storage of potential energy?

(A) a rushing river
(B) falling rain
(C) a lake held in by a dam
(D) steam released from a boiling kettle of water
(E) a melting ice cube

338. Work performed per unit of time is called

(A) energy
(B) potential energy
(C) power
(D) the law of conservation of energy
(E) electromagnetism

339. According to the law of conservation of energy,

(A) power is a measure of the conversion of energy
(B) work performed on an object results in a change in its energy
(C) nuclear fusion converts some matter into huge amounts of energy
(D) potential energy can be converted into kinetic energy
(E) energy cannot be created or destroyed, only converted into different forms

340. Using all of the energy she can get from eating a peanut butter and jelly sandwich, a woman takes as many buckets of water to the top of a hill as she can and pours them into a barrel there. She opens a valve at the bottom of the barrel, and all the water flows down a pipe running down the hill. The pipe ends halfway to the bottom, where the water strikes the paddles of a waterwheel, which turns in response. As the wheel turns, it coils a spring. Which of the following objects contained the greatest total energy?

(A) the water in the barrel at the top of the hill, before it flowed into the pipe
(B) the coiled spring, right when all the water had flowed from the pipe
(C) the water moving down the pipe, right when the barrel emptied
(D) the sandwich, before the woman took her first bite
(E) the waterwheel, when it first began to coil the spring

341. Which of the following heating methods involves the most efficient conversion from its naturally occurring form to heat?

(A) whale oil burned in an old-fashioned lantern
(B) sunlight through the window absorbed by the walls of a room
(C) heating oil in a house's furnace
(D) sunlight collected by a solar panel to run an electric heater
(E) gasoline running a generator that powers an electric heater

342. The two main elements that compose fossil fuels are

(A) oxygen and hydrogen
(B) oxygen and carbon
(C) carbon and hydrogen
(D) sulfur and carbon
(E) sulfur and hydrogen

343. Which of the following lists of energy sources is ordered from greatest to smallest percentage of total annual energy usage in the United States?

(A) nuclear fuels, fossil fuels, geothermal energy
(B) geothermal energy, nuclear fuels, fossil fuels
(C) fossil fuels, nuclear fuels, geothermal energy
(D) fossil fuels, wind, nuclear fuels
(E) wind, fossil fuels, nuclear fuels

344. Which three of the following are renewable energy sources?

I. biomass
II. natural gas
III. coal
IV. hydropower
V. oil
VI. solar power

(A) I, II, and VI
(B) I, IV, and VI
(C) II, IV, and VI
(D) II, III, and V
(E) III, IV, and V

345. All of the following are typically derived from fossil fuels EXCEPT

(A) geothermal power
(B) plastics
(C) organic chemicals
(D) U.S. electrical power
(E) automobile fuel

346. Most of the world's oil is located beneath countries in

(A) South America
(B) the Far East
(C) Europe
(D) the Middle East
(E) the South Pacific

347. Of oil drilled offshore in U.S. waters, about 90% is recovered from under

(A) the Gulf of Mexico
(B) Chesapeake Bay
(C) the Great Lakes
(D) the North Atlantic
(E) the Gulf of Alaska

348. The United States uses about what proportion of world oil production each year?

(A) 10%
(B) 25%
(C) 50%
(D) 75%
(E) 90%

349. Compared with the production cost of $7.50 to $10 per barrel of U.S. oil, Saudi Arabian oil is

(A) less than half as expensive to produce
(B) between half as expensive and the same cost to produce
(C) about the same cost to produce
(D) twice as expensive to produce
(E) four times as expensive to produce

350. Canada has large deposits—about three-fourths of the world's supply—of an unconventional oil known as

(A) bitumen, or tar sands
(B) shale oil, or kerogen oil
(C) biomass-based liquid supplies
(D) coal-based liquid supplies
(E) natural gas–based liquid supplies

351. Compared with conventional petroleum oil, the amount of carbon dioxide released by burning unconventional oils is

(A) about one-tenth as large
(B) about half as large
(C) about the same
(D) about twice as large
(E) four times as large

352. When energy extractors find it underground, "natural gas" is a mixture of gases, but it's mostly composed of

(A) butane
(B) propane
(C) methane
(D) ethane
(E) hydrogen sulfide

353. The one major pollutant released by burning natural gas is

(A) nitrous oxide
(B) hydrogen sulfide
(C) hydrogen cyanide
(D) carbon dioxide
(E) sulfur dioxide

354. Of the following lists of the world's energy supplies, which begins with the supply that is likely to last longest and ends with the supply that will likely run out soonest?

(A) solar, conventional oil, natural gas
(B) solar, natural gas, conventional oil
(C) coal, solar, conventional oil
(D) conventional oil, natural gas, coal
(E) biomass, conventional oil, natural gas

355. Before it is transported in pipelines or liquefied form, natural gas must be rid of poisonous

(A) nitrogen gas
(B) hydrogen sulfide gas
(C) butane gas
(D) methane gas
(E) ethane gas

356. Unconventional natural gas sources include

(A) oil shale
(B) tar sand
(C) methane hydrate
(D) anthracite
(E) lignite

357. Contour strip mining is a method specialized to extract coal in terrain that is

(A) hilly or mountainous
(B) generally flat
(C) underwater
(D) covered with swamps or marshland
(E) coastal

358. The most abundant fossil fuel is

(A) unconventional natural gas
(B) petroleum
(C) coal
(D) oil shale
(E) tar sand

359. The single country with the largest natural gas reserves is

(A) Qatar
(B) the United States
(C) Canada
(D) Russia
(E) Iran

360. When burned, which of the following fossil fuels produces the smallest amount of carbon dioxide per unit of energy released?

(A) gasoline
(B) coal
(C) natural gas
(D) bitumen
(E) oil shale

361. In the next 10 to 20 years, U.S. natural gas production is expected to increase due to greater exploitation of domestic

(A) shale deposits
(B) tar sands
(C) conventional natural gas deposits
(D) methane hydrate
(E) anthracite

362. Transporting natural gas can be expensive for which two of the following reasons?

I. It requires the building and maintenance of special pipelines.
II. It must be cooled to a very low temperature to liquefy for shipping.
III. It must undergo a hydraulic fracturing to prepare it for fuel use.
IV. It must be purified of radioactive contaminants before it can be transported.

(A) I and II
(B) I and III
(C) II and III
(D) II and IV
(E) III and IV

363. Of the following lists of coal types, which one begins with the type having the highest energy content and ends with the type having the lowest energy content?

(A) lignite, bituminous, anthracite
(B) bituminous, anthracite, lignite
(C) anthracite, bituminous, lignite
(D) lignite, anthracite, bituminous
(E) anthracite, lignite, bituminous

364. With about 25% of the world's proven reserves, the country with the largest coal supply is

(A) Russia
(B) China
(C) Bolivia
(D) the United States
(E) Australia

365. Fossil fuels typically found beneath a dome formation either on the earth's surface or under the seafloor include

I. natural gas
II. crude oil
III. coal

(A) I only
(B) II only
(C) III only
(D) I and II
(E) II and III

366. The Bog of Allen in Ireland is an ancient lake that gradually began filling with plant matter about 10,000 years ago and is now completely filled. As plant matter accumulated, it was pushed to the bottom of the lake and compressed, eventually undergoing slow chemical reactions that changed it into peat. For thousands of years, people living in the area have burned peat from the bog for fuel, and starting in the 1900s, peat production became an industrialized process. About 90% of the bog is now dry or has been removed for use as fuel. Other ancient wetlands have managed to maintain a thin layer of water for a very long time, allowing peat to accumulate for hundreds of thousands of years, due to gradual sinking of the land or gradual rising of the water level as plant matter accumulates. In these places, sediment occasionally washes over the peat to compress it further, and peat at the bottom of the heap slowly turns into coal. Peat has a carbon content of about 60% when dry, while anthracite coal has a carbon content of about 87%.

(A) Is peat a fossil fuel or a renewable fuel?
(B) How does the carbon content of bituminous coal compare to that of peat and anthracite coal?
(C) Is peat from the Bog of Allen likely to ever become coal?

367. Around the beginning of the 21st century, fossil fuel extraction companies began to discover huge reserves of natural gas trapped in shale deposits. Recently developed improvements in hydraulic fracturing technology have allowed gas extractors to remove the fossil fuel from the ground more efficiently than was previously possible, and in 2009, the Colorado School of Mines' Potential Gas Committee recognized a record increase in U.S. natural gas reserves—about 35% more natural gas had recently become available for extraction due to the recognition of unconventional gas from shale. However, hydraulic fracturing has a growing number of critics, who are concerned that carcinogenic and otherwise dangerous chemicals from fracturing fluid may contaminate underground water supplies, while natural gas may have already leaked into water wells in Dimock, Pennsylvania.

(A) How does hydraulic fracturing help to remove natural gas from shale deposits?
(B) Does unconventional natural gas from shale contribute to global warming more or less than conventional natural gas?
(C) How would hydraulic fracturing fluids or natural gas get from shale gas wells into drinking water? How deep are water wells, compared to the depth at which fracturing fluid is injected into shale deposits?

368. In the United States, energy experts have been contemplating the benefits of turning coal into a liquid fuel to replace petroleum-based fuels such as gasoline and diesel. According to the American Association for the Advancement of Science, the justification for coal-to-oil approaches is to prepare for higher oil prices and to reduce U.S. dependence on foreign petroleum. There are two basic ways to turn coal into a liquid fuel—indirect liquefaction and direct liquefaction. The first is a multistep process in which production facilities turn coal into a gas called syngas; they then remove impurities such as sulfur and refine it into diesel or gasoline fuels. According to AAAS, the full indirect liquefaction process is three to four times more expensive than processing the same amount of oil, and it releases a great deal of carbon dioxide even before the fuel is burned. When taking into account the cost of new infrastructure and the capturing of carbon dioxide released during refinement, indirect liquefaction becomes more expensive still. In the second major procedure, direct liquefaction, processing facilities use high-temperature chemical reactions to produce a liquid fuel. This fuel is currently used in China, but it contains many impurities that prevent complete combustion, resulting in emissions of unburned fuel as well as other impurities that make it illegal to use in the United States. As a result, its equivalent per-barrel price is even higher than that of fuel produced by indirect liquefaction.

(A) How does the second law of thermodynamics increase the price of energy from coal-derived liquid fuel?

(B) What factors might encourage U.S. adoption of coal-derived liquid fuel?

(C) What are the likely negative effects of the incomplete combustion of fuel derived from direct coal liquefaction?

(D) Even if coal-derived liquid fuel is highly refined and the process is made emission-free, what major environmental risks would it still produce?

Nuclear Energy

369. The uranium oxide pellets used in typical nuclear reactors are primarily composed of the nonfissionable isotope

(A) uranium-235
(B) uranium-238
(C) uranium-258
(D) uranium-255
(E) uranium-245

370. Inside a reactor, a neutron strikes a uranium atom nucleus, which splits, releasing a great deal of energy and sending fragments of the nucleus speeding away in different directions. This single sequence of events is best described as an example of

(A) fusion
(B) fission
(C) enrichment
(D) meltdown
(E) beta radiation

371. In addition to cooling a nuclear reactor's core, a fluid known as *coolant* serves the additional purpose of

(A) transferring heat out of the core for electricity generation
(B) keeping radioactive gases from escaping into the environment
(C) keeping fuel from chemically reacting with the cooling rods
(D) adding neutrons to nonfissionable material to create new nuclear fuel
(E) absorbing neutrons to control the rate of fusion

372. After operating for up to 60 years, conventional nuclear plants must be either renovated or decommissioned for which two main reasons?

I. Many parts become radioactive.
II. Spent fuel rods cannot be removed from the core.
III. Many parts become brittle or corroded.
IV. Pressure inside the core increases until it reaches the facility's limit.

(A) I and II
(B) II and III
(C) I and III
(D) II and IV
(E) I and IV

373. Fissionable uranium-235 contains a total of 235

(A) protons
(B) neutrons
(C) electrons
(D) Higgs bosons
(E) protons and neutrons

374. Of the following lists of radioactive decay products, which is arranged from heaviest to lightest?

(A) gamma particles, alpha particles, beta particles
(B) beta particles, alpha particles, gamma particles
(C) gamma particles, beta particles, alpha particles
(D) alpha particles, beta particles, gamma particles
(E) alpha particles, gamma particles, beta particles

375. Whether or not an atom's nucleus is unstable depends mostly on its number of

(A) protons
(B) electrons
(C) neutrons
(D) quarks
(E) pi orbitals

376. The half-life of element X is 40 years. Starting from a mass of 50 grams, how much element X will have undergone radioactive decay in 120 years?

(A) 50 grams
(B) 43.75 grams
(C) 25 grams
(D) 12.5 grams
(E) 6.25 grams

377. Radioactive materials are dangerous to organisms such as humans because

(A) they neutralize the electrical charges of biological molecules, rendering them inert
(B) cells will substitute radioactive elements for oxygen and suffocate
(C) they cause proteins to replicate themselves uncontrollably
(D) they release particles that ionize biological molecules, splitting them
(E) they absorb crucial electrolytes, shutting down ion transfer across cell membranes

378. A nuclear reactor that makes more fissionable fuel is specifically known as a

(A) fusion reactor
(B) tokamak reactor
(C) moderator reactor
(D) breeder reactor
(E) light-water reactor

379. When spent nuclear fuel rods are stored temporarily at a nuclear reactor, they are often put into pools of boron-treated water to prevent them from

(A) undergoing an uncontrollable chain reaction resulting in a nuclear explosion
(B) turning to difficult-to-manage radioactive dust after further decay
(C) spreading throughout the facility as they melt into liquid form
(D) heating up, catching fire, and releasing contaminants into the environment
(E) combining with nitrogen in the air to become a high-pressure radioactive gas

380. The open nuclear fuel cycle does NOT involve

(A) reprocessing spent fuel into usable fissionable material
(B) burying radioactive wastes underground for thousands of years
(C) mining uranium-containing ore from the earth
(D) decommissioning old reactors
(E) temporarily storing spent fuel in dry casks or pools of water

381. To safeguard the environment and people living in it, high-level radioactive waste from fission reactors must be stored for

(A) 1 to 10 years
(B) 50 to 5,000 years
(C) 10,000 to 250,000 years
(D) 1 million to 10 million years
(E) 500 million to 1 billion years

382. What proportion of the United States' high-level radioactive waste is stored in long-term underground facilities?

(A) all of it
(B) 75%
(C) 50%
(D) 25%
(E) none of it

383. Transuranium elements can be obtained only by

(A) mining deposits located deep under the ocean
(B) refining petroleum using high-pressure processes
(C) bombarding heavy elements with neutrons
(D) treating uranium ores with corrosive chemicals
(E) gathering and refining uranium decay products

384. Which of the following is NOT a drawback of nuclear energy?

(A) highly dangerous waste products
(B) potential for catastrophic plant accident
(C) cannot compete economically without government subsidies
(D) high air pollution from normal operations
(E) plants and waste storage facilities seen as terrorist targets

385. Which three of the following are components of the open nuclear fuel cycle?

I. decommissioning of old power plants
II. mining uranium ore
III. generating electricity with a steam-driven turbine
IV. reprocessing of radioactive waste into fuel
V. pumping of coolant through the core
VI. temporary storage of high-level radioactive waste

(A) I, II, and III
(B) I, II, and VI
(C) II, III, and IV
(D) III, IV, and V
(E) II, IV, and VI

386. Most of the world's nuclear reactors and all of the United States' nuclear reactors are

(A) light-water reactors
(B) fusion reactors
(C) pebble-bed modular reactors
(D) high-temperature gas-cooled reactors
(E) breeder reactors

387. In the United States, the government agency currently charged with overseeing the safety and licensing of nuclear reactors and their fuel is known as the

(A) U.S. Nuclear Regulatory Commission
(B) U.S. Atomic Energy Commission
(C) U.S. Union of Concerned Scientists
(D) U.S. Energy Research and Development Administration
(E) U.S. National Nuclear Security Administration

388. The 1986 Chernobyl nuclear power plant accident occurred in

(A) Russia
(B) Ukraine
(C) Lithuania
(D) France
(E) Belarus

389. Coal compares favorably to nuclear energy in few areas, including

(A) land disruption due to mining
(B) air pollution
(C) contribution to acid rain
(D) difficulty of plant construction and maintenance
(E) contribution to global warming

390. Among the following U.S. regions, which has the highest concentration of nuclear power plants?

(A) Mountain West
(B) Southwest
(C) Midwest
(D) Gulf Coast
(E) Pacific Coast

391. The planned Yucca Mountain long-term, underground storage facility for high-level radioactive waste is located in the state of

(A) Wyoming
(B) Nevada
(C) Utah
(D) Idaho
(E) Colorado

392. In order for uranium ore to be enriched for use as nuclear power plant fuel, it is first chemically converted to

(A) uranium hexafluoride
(B) uranium hexachloride
(C) uranium hexabromide
(D) plutonium-239
(E) uranium hexaiodide

393. Pure uranium ore, or "yellowcake," is composed of the element uranium chemically bonded to the element

(A) fluorine
(B) chlorine
(C) oxygen
(D) boron
(E) hydrogen

394. Of the following power plant components, which one is found ONLY in nuclear power plants?

(A) turbine
(B) cooling tower
(C) control rod
(D) condenser
(E) water pump

395. In the closed nuclear fuel cycle, decommissioned reactors are

(A) processed into fissionable fuel
(B) reused in the construction of new power plants
(C) simply discarded in a landfill, since they are not radioactive
(D) buried or otherwise disposed of safely, since they are radioactive
(E) reduced to ash through incineration

396. After the 2011 earthquake and tsunami that devastated some of Japan's east coast, the Fukushima Daiichi power plant experienced an uncontrolled fission reaction when coolant stopped circulating around fuel rods, and emergency control measures did not adequately halt the process. Thus, the plant experienced a partial

(A) enrichment
(B) meltdown
(C) conversion
(D) fission explosion
(E) nuclear fusion reaction

397. Fusion reactors are not used for energy production because they

(A) produce dangerous amounts of radiation
(B) can explode with the force of a hydrogen bomb in an accident
(C) produce a large amount of high-level radioactive waste compared with fission reactors
(D) have not yet produced more energy from fuel than they consume
(E) require fuel that is not economically effective to produce

398. A fusion reactor–based power plant would make electricity from heat released by the fusion of nuclei of light elements such as

(A) nitrogen
(B) hydrogen
(C) lithium
(D) beryllium
(E) boron

399. In 2003, the Paks Nuclear Power Plant near Paks, Hungary, experienced a serious incident that could have been much worse. While a special machine was being used to clean fuel rods at the bottom of a special water pool, inadequate water circulation allowed some rods to heat up, probably cracking them a little, and the pool's water began to register an increased level of radioactivity. When workers opened the cleaning device underwater, a sudden change in temperature damaged the rods' outside coating enough to allow uranium fuel pellets to spill into the bottom of the cleaning tank, with some of the elements piling together—a situation that caused inspectors to immediately raise their assessment of the incident's seriousness. Power plant workers poured boric acid into the water, and they eventually cleaned up the mess.

(A) Even if the rods had not released uranium pellets, what might have happened had the rods been allowed to heat up further?
(B) Why would inspectors find a pile of uranium pellets so concerning?
(C) Why would workers pour boric acid into the pool? What effect would it have?

400. A company called General Fusion aims to build a working model of a novel type of fusion reactor by 2013. The reactor involves a spinning sphere of liquid metal—lead and lithium—with a hollow internal cylinder maintained by the constant rotation. Inside the hollow chamber is a magnetically suspended deuterium-tritium fuel, which is compressed about one time every second by pistons hammering on the outside layer of the spinning metal sphere. The resulting compression wave is supposed to cause brief nuclear fusion. This reaction releases neutrons that are slowed down by the metal, which heats up. Some hot metal is extracted to create electricity.

(A) What are deuterium and tritium?
(B) Why is the plasma magnetically suspended rather than kept in a solid container of some sort?
(C) How does General Fusion probably plan to generate electricity using hot metal?

401. Beyond low-earth orbit, space is awash in dangerous radiation, especially fast-moving protons and electrons. Here on earth, and in low orbit, we're protected mostly by the planet's magnetic field, which reflects or redirects a lot of radiation. NASA scientists are exploring the idea of using a similar concept to protect astronauts in bases on the moon by suspending large positively and negatively charged balloons high above places where astronauts would live and work. A negatively charged balloon would reflect negatively charged particles, while a positively charged balloon would reflect positively charged particles.

(A) What is alpha radiation?
(B) What is the name of radiation composed of fast-moving electrons?
(C) What sort of radiation is NOT blocked by positively OR negatively charged fields?

Alternative and Renewable Energies

402. Of the following energy sources, which one is a "secondary source" that must be produced by humans using another energy source that may be nonrenewable?

(A) solar
(B) hydrogen
(C) biomass
(D) geothermal
(E) wind

403. Dams that allow water to flow through at a controlled rate can generate electricity if the water turns a turbine on its way out. This kind of power plant produces

(A) dam power
(B) geothermal power
(C) kinetic power
(D) tidal power
(E) hydroelectric power

404. Which of the following is often a disadvantage of damming rivers to generate electricity?

(A) low-efficiency conversion of water kinetic energy to electricity
(B) short power-plant life, requiring frequent decommissioning
(C) negative environmental impact from flooding behind a dam and decreased fertilizer flow through a whole river
(D) high carbon dioxide emissions from electricity-generating equipment
(E) water behind a dam not suitable for agricultural purposes

405. Wafers of layered semiconductor sheets that produce electrical current when exposed to light are technically known as

(A) photovoltaic cells
(B) thermal cells
(C) visible spectrum cells
(D) passive solar cells
(E) semiconductor cells

406. The Staebler-Wronski effect describes

(A) the drop in efficiency of certain silicon solar cells exposed to intense light
(B) an increase in solar-cell efficiency related to lower operating temperatures
(C) the decrease in efficiency of wind turbines in winds faster than 25 kilometers per hour
(D) the backward flow of electrical current in solar cells lacking any light
(E) the loss of some energy in the form of waste heat as energy is converted from one form to another

407. Which two of the following alternative energy sources do NOT ultimately depend on energy from the sun?

I. wind
II. dammed rivers
III. dammed ocean coves
IV. biomass
V. geothermal heat

(A) I and II
(B) II and III
(C) III and IV
(D) III and V
(E) IV and V

408. One of the main problems with switching to ethanol as a fuel in the United States is that

(A) it takes a great deal of energy to produce ethanol from corn, the major crop used in its production
(B) automobile engines have not yet been developed to use ethanol
(C) ethanol cannot be made in a renewable way
(D) burning ethanol releases large amounts of sulfur into the atmosphere
(E) producing ethanol requires the use of highly toxic chemicals

409. Of the following alternative energy sources, which has the highest energy return on energy invested?

(A) photovoltaic solar
(B) hydroelectric
(C) geothermal
(D) biodiesel
(E) wind

410. Possible alternative automobile fuels include all of the following EXCEPT

(A) hydrogen
(B) ethanol
(C) biomass
(D) biodiesel
(E) methanol

411. In the United States, mass production of which of the following alternative energy sources could cause an increase in food prices?

(A) geothermal
(B) wind
(C) ethanol
(D) solar
(E) hydroelectric

412. Proposed methods for using geothermal energy include

(A) running turbines directly with hot gases expelled through volcanic vents
(B) channeling molten magma from live volcanoes to power plants to run steam turbines
(C) capturing hydrogen sulfide gas for use as a flammable fuel
(D) heating a home's water by piping it underground and back
(E) using the magnetic fields produced by churning magma in the mantle to create electricity by induction

413. Wind turbines can produce more power when they

(A) generate direct current rather than alternating current
(B) are constructed with longer blades
(C) are mounted higher off the ground
(D) have five or more blades each
(E) include small electric motors to aid spinning

414. All of the following are examples of wind power's disadvantages EXCEPT

(A) wind power can be inconsistent
(B) wind farms require a large amount of space
(C) wind turbines make noise
(D) wind turbines generate direct current that must be converted to alternating current
(E) wind turbines can kill birds in flight

415. Which three of the following energy sources are NOT necessarily sustainable?

I. wind
II. biomass
III. hydrogen
IV. solar
V. geothermal
VI. ethanol

(A) I, II, and III
(B) II, III, and VI
(C) III, IV, and V
(D) III, V, and VI
(E) IV, V, and VI

416. While methanol and ethanol are very similar, the major structural difference between the two chemicals is that

(A) methanol has a phosphorus backbone, while ethanol has a carbon backbone
(B) methanol has one carbon atom surrounded by hydrogen, while ethanol has two
(C) methanol has one hydroxyl (-OH) group, while ethanol has two
(D) methanol is not an alcohol, while ethanol is
(E) methanol has two carbon atoms surrounded by hydrogen, while ethanol has one

417. A hydroelectric facility that takes in only some of a river's water—rather than damming the entire river—is known as

(A) an impoundment plant
(B) a diversion plant
(C) a pumped storage plant
(D) an irrigation and generation plant
(E) a micro hydropower plant

418. Which two of the following alternative energy sources do NOT require the use of a turbine to create electricity?

I. solar
II. wind
III. geothermal
IV. hydropower
V. ethanol

(A) I and II
(B) I and V
(C) II and III
(D) II and IV
(E) III and V

419. Burning hydrogen releases energy in the form of heat and

(A) carbon dioxide
(B) hydrogen sulfide
(C) water
(D) oxygen gas
(E) hydrogen peroxide

420. Burning pure ethanol releases mainly

(A) carbon dioxide and water
(B) oxygen and hydrogen gases
(C) carbon monoxide and nitrous oxide
(D) hydrogen sulfide and hydrogen peroxide
(E) sulfur dioxide and diethyl ether

421. Processing biomass with bacteria can produce which three of the following fuels?

I. methane
II. methanol
III. ethanol
IV. hydrogen
V. butane
VI. charcoal

(A) I, II, and III
(B) I, III, and IV
(C) II, III, and IV
(D) II, IV, and V
(E) III, IV, and VI

422. Two countries together produce and consume the majority of the world's ethanol fuel. They are

(A) Russia and Nigeria
(B) Canada and the Netherlands
(C) Iceland and New Zealand
(D) Argentina and China
(E) the United States and Brazil

423. Which of the following is the LEAST efficient way to use energy stored in biomass?

(A) directly burning it to produce heat in an open fire
(B) converting it to wood gas to burn for heat
(C) converting it to ethanol for use as an automobile fuel
(D) burning it to run a steam turbine for electricity to make hydrogen from water for automobile fuel
(E) using it in bacterial processors to make methane for use in a steam turbine–based electricity-generating plant

424. When ethanol is used as a fuel in the United States, it is mostly in the form of

(A) a pure liquid for use in special automobiles
(B) a pressurized gas that is usually burned for its heat
(C) a liquid mixed with gasoline for use in ordinary automobiles
(D) a pressurized gas that is used only in specially outfitted automobiles
(E) a solid "brick" that is usually burned for its heat

425. Compared to ordinary gasoline, one disadvantage of ethanol is its

(A) higher carbon dioxide emissions
(B) lower fuel efficiency per unit volume
(C) higher carbon monoxide emissions
(D) lower torque (rotational force) output
(E) higher sulfur emissions

426. Of the following materials, which CANNOT be used as a biomass fuel?

(A) charcoal
(B) granite
(C) animal dung
(D) cardboard
(E) tree bark

427. Of the following countries' capital cities, which one uses geothermal heat to warm the great majority of its buildings?

(A) Washington, D.C.
(B) Wellington, New Zealand
(C) Reykjavik, Iceland
(D) Moscow, Russia
(E) Kiev, Ukraine

428. One disadvantage to tapping dry- or wet-steam geothermal resources is that

(A) drilling that deeply runs the risk of provoking a volcanic eruption
(B) it is possible to deplete the heat in some underground reservoirs for a period of time
(C) releasing underground pressure can cause the overlying ground to sink, cracking buildings and streets
(D) drilling too close to a geologically active fault can increase the frequency of earthquakes
(E) water from deep in the earth's crust is highly corrosive to metal pipes and machines, making maintenance very expensive

429. Most U.S. states require electric utilities to provide a certain share of electricity from renewable sources. They meet this goal by buying certificates from suppliers of electricity from sustainable sources. These certificates represent 1,000 kilowatt-hours of electricity, and they are called

(A) sustainable electricity certificates, or green certs
(B) renewable energy certificates, or green tags
(C) replenishable source certificates, or green credentials
(D) environmentally sound certificates, or green records
(E) inexhaustible supply certificates, or green papers

430. Which of the following has NOT been proposed as a method for storing hydrogen fuel?

(A) cooling it to liquid form
(B) compressing it as a gas
(C) absorbing it with metal hydride compounds
(D) absorbing it in activated charcoal
(E) cooling it to solid form

431. By far the largest alternative energy source by total energy production worldwide is

(A) hydropower
(B) wind
(C) biomass
(D) geothermal
(E) solar

432. Taking advantage of its sunny weather, Spain is investing heavily in solar power. One project is the PS10 solar power plant near Seville, the first "commercial concentrating solar power tower" in the world. PS10 generates 11 megawatts of electricity using 624 swiveling mirrors—each 120 square meters—to reflect sunlight onto a collector at the top of a 115-meter tower, where it heats water to run a steam-driven turbine. In 2011 dollars, the plant cost about $50 million to build, and it's part of a larger, 300-megawatt complex that is expected to cost a total of about $1.7 billion. At about $5.67 per megawatt, the Spanish solar power complex is more expensive than the estimated $4-per-megawatt construction cost that Alliant Energy estimated for a new coal-based power plant near Portage, Wisconsin, in 2008. The Wisconsin project's original estimated price tag was about $2.83 per megawatt, but construction costs rose sharply thereafter. The 2007 price of electricity from PS10 was about three times that of electricity from other sources.

(A) What type of solar power does the PS10 plant use to generate electricity?
(B) Which aspects of the operation of PS10 have cost advantages over a coal-fired power plant?
(C) What environmental advantages might PS10 have over a coal-fired plant?

433. When Hoover Dam opened in 1936, it was the largest concrete structure in the world. Beginning in 1931, thousands of workers contributed to the project—more than 5,200 were working when the project was at its employment peak. Located at the border of Nevada and Arizona, Hoover Dam generates 2,080 megawatts for utilities in those two states and California. It's the single reason for the existence of the Lake Mead reservoir, which lies immediately upstream, and it controls flooding of the Colorado River below it.

(A) What sort of hydroelectric power plant is Hoover Dam? What is its power classification?
(B) Would electricity from Hoover Dam be fairly represented as renewable energy?
(C) What are Hoover Dam's negative environmental effects?

434. The largest tidal power plant in the world is France's La Rance tidal power station. Finished in 1966, it was also the first such plant, and today it generates 240 megawatts of power. The plant works by using a small dam—called a *barrage*—to close a coastal estuary from the ocean. When tides come in, water on the ocean side of the barrage rises, and it is allowed to flow through the barrage only at points where it can turn an electricity-generating turbine. When the tide recedes, the reverse process occurs, and water flowing out to the ocean must again pass through turbines to get through the barrage. The project has resulted in silting of the estuary, and some species, such as the sand eel, have suffered as a result.

(A) Is the La Rance power plant more or less reliant upon rainwater and melting snow than most hydroelectric power plants?

(B) Where does tidal energy ultimately come from?

(C) What are the likely negative environmental effects of a barrage-type tidal power plant?

Pollution Types

435. Smog levels are usually highest in

(A) fall
(B) winter
(C) spring
(D) summer
(E) same all year round

436. Sulfur dioxide, nitrogen oxides, and particulate emissions were limited by the

(A) Clean Air Interstate Rule (CAIR)
(B) Corporate Average Fuel Economy (CAFE)
(C) Energy Policy and Conservation Act of 1975
(D) Clean Air Act
(E) Comprehensive Environmental Response, Compensation, and Liability Act (CERCLA)

437. What percentage of the streams in the United States are affected by agriculture?

(A) 25%
(B) 60%
(C) 30%
(D) 10%
(E) 5%

438. Municipal water pollution comes from

(A) chemical pesticides
(B) oil operations
(C) wastewater
(D) fertilizers
(E) mining

439. Red tide blooms are caused by

(A) cryptosporidium
(B) dead zones
(C) El Niño
(D) oil spills
(E) storm runoff

440. Which of the following is the most ideal situation for a stream in terms of amount of oxygen to support animals?

(A) cold temperature, fast-moving, wide, lots of native vegetation
(B) warm temperature, slow-moving, narrow, no vegetation
(C) cold temperature, slow-moving, narrow, lots of vegetation
(D) warm temperature, fast-moving, wide, no vegetation
(E) none of the above affect dissolved oxygen amounts

441. The main method of water disinfection in major cities is

(A) ozone oxidation
(B) ultraviolet light
(C) turbidity
(D) chlorination
(E) reaeration capacity

442. Nonpoint source contamination of water includes all of the following EXCEPT

(A) rainfall
(B) pesticide residues in soil
(C) water from waste treatment plants
(D) emissions from automobiles
(E) fertilizers

443. Which nutrient becomes a pollutant and is found in detergents and fertilizer?

(A) nitrogen
(B) phosphorus
(C) hazardous waste
(D) pathogens
(E) organic matter

444. Acid rain affects trees by

I. damaging the protective coating on leaves
II. leaching nutrients from the soil and away from trees
III. stunting growth

(A) I only
(B) II only
(C) I and II only
(D) II and III only
(E) I, II, and III

445. An agreement between Canada and the United States to limit acid-emitting sources is called the

(A) Superfund
(B) OSHA
(C) CERCLA
(D) Clean Water Act
(E) Air Quality Accord

446. Power plants, industry, mining, and medical tracers are all sources of what type of pollution?

(A) pathogens
(B) oil spills
(C) radioactive waste
(D) thermal pollution
(E) acid emission

447. When was Earth Day first established?

(A) April 22, 1970
(B) March 22, 1972
(C) April 1, 1975
(D) March 22, 1970
(E) April 20, 1965

448. Chlorofluorocarbons are pollutants because they

(A) turn lakes acidic
(B) break down the ozone layer
(C) contribute to global warming
(D) decrease the biochemical oxygen demand
(E) both B and C

449. Which by-product of manufacturing is slowly released from sediments into water and is biomagnified in the food chain, causing numerous health problems?

(A) sodium
(B) nitrates
(C) chlorine
(D) methyl mercury
(E) PVCs

450. The Plastic Pollution Control Act of 1988

(A) bans the production of nonbiodegradable plastic
(B) makes it illegal for boats to discard plastic trash into the ocean
(C) prohibits the dumping of sludge
(D) sets limits on manufacturing of plastic
(E) limits emissions of plastic manufacturing

451. When different pollutants combine to have a greater effect than each individual pollutant, this is known as

(A) bioaccumulation
(B) biomagnification
(C) synergism
(D) contamination
(E) toxicity

452. A dangerous man-made chemical produced as a by-product of herbicide manufacturing that causes skin and liver problems, cancer, and hormone disruption is

(A) dioxin
(B) PCB
(C) lead
(D) sulfur dioxide
(E) asbestos

453. Volcanoes, fossil fuel combustion, and fires contribute to pollution in the form of

(A) synthetic compounds
(B) DDT
(C) EMFs
(D) sodium chloride
(E) particulates

454. Carbon monoxides, VOCs, hydrocarbons, nitrogen oxides, peroxyacetyl nitrate, benzene, and lead are all pollutants produced by

(A) indoor air pollution
(B) gasoline-powered vehicles
(C) mining
(D) manufacturing
(E) biomass burning

455. Which of the following laws regulates new chemicals as well as other chemicals that were in existence before the enactment of this law?

(A) Safe Drinking Water and Toxic Enforcement Act
(B) Toxic Substances Control Act (TSCA)
(C) Green taxes
(D) Federal Hazardous Substances Act
(E) Food, Drug, and Cosmetic Act

456. Alcohol, thalidomide, radiation, and rubella are all known

(A) neoplasms
(B) carcinogens
(C) teratogens
(D) PCBs
(E) DDT

457. Pollution that comes from human sources such as cars, power plants, and industry is known as

(A) anthropogenic
(B) natural
(C) photochemical
(D) secondary
(E) volatile

458. Noise pollution is measured in what units?

(A) hertz (Hz)
(B) milligrams (mg)
(C) centimeters (cm)
(D) decibels (dB)
(E) watts (W)

459. Which pollutant can be found indoors in carpet, furniture, and plywood and is dangerous to human health?

(A) radon
(B) asbestos
(C) carbon monoxide
(D) VOCs
(E) formaldehyde

460. One pollution control measure in which companies can sell "credits" for emissions is known as

(A) deposition credits
(B) tradable license
(C) deposition permits
(D) tradable permits
(E) pollution penalties

461. Which of the following is the most dangerous indoor air pollutant, causing more deaths than any other?

(A) radon
(B) secondhand smoke
(C) formaldehyde
(D) asbestos
(E) dust mites

462. A chemical that causes respiratory problems as well as damage to the environment and is a part of photochemical smog is

(A) ozone
(B) arsenic
(C) chlorine
(D) asbestos
(E) methane

463. Which organisms are most affected by air pollution because they obtain their nutrients from the air?

(A) frogs
(B) trees
(C) lichens
(D) birds
(E) bacteria

464. An example of a pollutant that occurs only naturally is

(A) formaldehyde
(B) nitrous oxide
(C) sulfur dioxide
(D) asbestos
(E) radon

465. Not all chemicals are equal when it comes to being dangerous. To decide if something is a pollutant to the environment, both the chemical itself and the amount must be considered to determine the threat to human and environmental health.

(A) Distinguish between the LD_{50} and TD_{50} of a pollutant.
(B) Explain why the Clean Water Act may have no threshold for a pollutant.
(C) Discuss an ecological gradient near a smokestack that emits sulfur dioxide in high quantities.

466. Mercury is one type of toxic inorganic pollutant that can affect living things. Other inorganic pollutants include other metals, salts, and acids.

(A) What are some potential sources of mercury?
(B) What are some ways that mercury enters the human body?
(C) List the effects mercury can have on human health.
(D) Using the ocean as an example, explain how mercury is biomagnified from one species to the next using specific examples.

467. Lead poisoning is particularly severe because it causes so many serious health problems, especially in children. This problem seems to occur at a higher incidence rate in African-American households at lower socioeconomic levels.

(A) Why does this pollutant disproportionately affect those in poverty?
(B) What are some ways to prevent lead poisoning?
(C) How are people in nonindustrialized countries exposed to lead, and how are they affected by it?

Global Change and Economics

468. Climate refers to

(A) conditions in the atmosphere that occur for weeks at a time in an area
(B) conditions in the atmosphere that occur for years at a time in an area
(C) the weather in the thermosphere
(D) the chemical elements in the atmosphere
(E) the amount of sunlight that reaches an area

469. The levels of carbon dioxide throughout the history of the earth can best be measured using

(A) ice cores
(B) fossils
(C) historical records
(D) snow
(E) rocks

470. Approximately how much of the sunlight reaching earth is reflected back into space?

(A) ¾
(B) ½
(C) ⅓
(D) ⅛
(E) ⅔

471. Deforestation, burning of fossil fuels, and other activities that release greenhouse gases all contribute to

(A) solar emissions
(B) changing ocean currents
(C) decreasing sea levels
(D) global warming
(E) sedimentation

472. Which of the following is NOT a source of methane?

(A) cows
(B) landfills
(C) forest fires
(D) dead plants
(E) aerosols

473. Which of these does not have a natural source?

(A) carbon dioxide
(B) CFCs
(C) nitrous oxide
(D) vapor
(E) hydrogen sulfide

474. What process do CFCs, chlorine, and bromine have in common that leads to destruction of the ozone layer?

(A) polymerization
(B) deposition
(C) photodissociation
(D) buffering
(E) biological control

475. The breakdown of the ozone layer is dangerous because it absorbs less

(A) UV radiation
(B) CFCs
(C) acid rain
(D) water vapor
(E) nitrogen

476. Rising sea levels that result from global warming are caused by

I. melting of glaciers
II. expanding oceans due to warmer temperatures
III. more rainfall

(A) I only
(B) II only
(C) I and II only
(D) II and III only
(E) I, II, and III

477. El Niño contributes to temporary global warming because it

(A) increases solar radiation
(B) increases deforestation
(C) causes rising sea levels
(D) increases electromagnetic radiation
(E) increases heat in the atmosphere from warm ocean waters

478. Which type of organisms would thrive in global warming conditions?

(A) small mammals
(B) butterflies
(C) migratory birds
(D) infectious insects
(E) grasses

479. Healthy oceans and forests

(A) are a carbon dioxide source
(B) are a carbon dioxide sink
(C) are an example of positive feedback
(D) contribute most of the greenhouse gases
(E) increase atmospheric carbon dioxide

480. What percentage of the world's carbon dioxide comes from the United States?

(A) 90%
(B) 5%
(C) 20%
(D) 60%
(E) 40%

481. Which fossil fuel releases the LEAST amount of carbon into the atmosphere?

(A) natural gas
(B) coal
(C) oil
(D) biomass
(E) timber

482. When energy from the sun gets absorbed by the ocean instead of being reflected by ice, this is known as

(A) El Niño
(B) polar amplification
(C) anthropogenic forcing
(D) solar radiation
(E) polar radiation

483. Which of the following requires the most energy to produce?

(A) recycled glass bottle
(B) new glass bottle
(C) recycled aluminum can
(D) new aluminum can
(E) new steel can

484. When a forest is destroyed for lumber, the loss of revenue from tourists not visiting that area is known as a(n)

(A) risk value
(B) cost mitigation
(C) marginal cost
(D) direct cost
(E) externality

485. The United States has mostly what type of economy?

(A) command economy
(B) market economy
(C) ecological economy
(D) sustainable economy
(E) free market system economy

486. Social pressure, laws and regulations, taxes, and licensing are all ways to make sure a group of people reach a common environmental goal. These methods are known as

(A) risk-benefit analyses
(B) supply and demand
(C) policy instruments
(D) economic analyses
(E) tragedy of the commons

487. The public relations problems that British Petroleum (BP) experienced after the Deepwater Horizon oil spill are known as

(A) direct costs
(B) supply costs
(C) equilibrium points
(D) repercussion costs
(E) indirect costs

488. An economic measure created by Herman Daly that takes into account environmental and safety issues when calculating human progress is the

(A) gross national product (GNP)
(B) gross domestic product (GDP)
(C) Index of Sustainable Economic Welfare (ISEW)
(D) Net Economic Welfare (NEW)
(E) Costs to Control Pollution (CCP)

489. The cost of reducing one additional unit of pollutant is the

(A) marginal cost
(B) pollution cost
(C) commons cost
(D) CCP cost
(E) public good cost

490. Which of the following has the highest risk of death?

(A) biking
(B) exposure to air pollution
(C) air travel
(D) football
(E) being a police officer

491. An example of an effective way that developed nations can assist developing nations is

(A) building dams
(B) clearing forests for agricultural use
(C) providing tools such as hand pumps for retrieving well water
(D) giving money to authoritarian governments
(E) building roads to remote regions

492. The amount of carbon dioxide that was in the atmosphere hundreds of years ago can be measured using

(A) fossils
(B) air samples
(C) ocean samples
(D) glacial ice cores
(E) tree sap

493. An example of a catalytic chain reaction is seen in

(A) formation of sulfur dioxides
(B) breakdown of DDT
(C) carbon sequestration
(D) increase of carbon dioxide
(E) ozone depletion

494. Nitrogen oxides make up what percentage of U.S. emissions of greenhouse gases?

(A) 35%
(B) 14%
(C) 60%
(D) 72%
(E) 6%

495. How much higher is the amount of carbon dioxide in the atmosphere today compared to the mid-1700s?

(A) 80%
(B) 50%
(C) 145%
(D) 10%
(E) 30%

496. The willingness of a businesswoman to postpone company profits by first implementing energy-efficient measures for increased profits in the future is known as

(A) time preference
(B) a discount rate
(C) sustainable profits
(D) cost-benefit analysis
(E) opportunity cost

497. Aesthetics and public service functions of the environment are also known as

(A) direct costs
(B) development incentives
(C) sustainable practices
(D) environmental intangibles
(E) certified sustainability

498. Global warming is a complex phenomenon that is influenced by many variables, including solar radiation, volcanic eruptions, human contributions to greenhouse gases, and weather system changes. The extent to which this warming occurs depends on how these variables interact with each other.

(A) Explain how each of the variables above can affect global warming.
(B) Give an example of a negative feedback cycle that occurs with one or more of these variables.
(C) Give an example of a positive feedback cycle that occurs with one or more of these variables.

499. Some people are not concerned about global warming because they are unaware of the direct impact it would have on them, their children, and their grandchildren. Suppose you have family members in different areas of our nation: Uncle Joe in the Midwest, Aunt Susanna in coastal Louisiana, and Grandma Betty in New Jersey. Write one letter to all of your family members explaining the potential changes they could expect to see in their lifetime due to global warming and, in particular, explaining how each of the three relatives mentioned might be affected.

500. In the wake of an oil spill, you are hired by the oil company to determine how to structure the pollution budget for the next oil-well drilling. You have a limited budget that must include the cost to prevent pollution as well as funds for cleanup in the event of an accident. Use the graph in Figure 15.1 to answer the following questions:

(A) Which line on the graph represents the cost of preventative measures?
(B) Which line represents the costs for cleanup?
(C) What does point A represent? Explain how you would use the information from point A to determine your budget limitations.

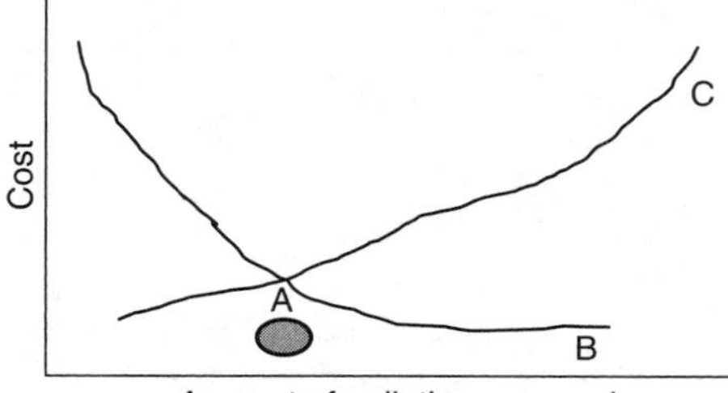

Figure 15.1

ANSWERS

Chapter 1: Earth Science

1. (C) Convection currents of magma flow slowly through the mantle. The lithosphere, core, and crust are all solid, while the hydrosphere consists largely of the earth's oceans.

2. (A) Radioactive isotopes in the earth's interior release high-energy particles into the surrounding rock, which warms it from the inside. Another major source of warming is the tidal pull of the moon.

3. (D) *Convection* refers to the circular warming-cooling motion of any heated liquid or gas.

4. (D) Transform faults slide against each other.

5. (C) Epoch, period, era, eon is the correct order.

6. (B) Hutton's principle of uniformitarianism posits that current geological features formed gradually over millions—and billions—of years.

7. (E) Divergent faults pull away from each other, as magma slowly wells up from the mantle to fill the gap, forming new crust.

8. (A) Tectonic plates push against each other until enough pressure causes them to slip and grind violently against one another in the form of an earthquake.

9. (E) Sedimentary rock is formed from tiny pieces of debris—including debris from other rock types—settling together and eventually adhering together to form a solid mass.

10. (C) As a logarithmic scale, the Richter scale's earthquake-magnitude increments each indicate a tenfold increase from the previous increment.

11. (E) Some volcanoes form at hot spots in tectonic plates, but the majority form around the edges, such as in the Pacific Ocean's Ring of Fire.

12. (D) As one plate is pushed into the earth's interior by another plate, it can form a giant trench that runs along the fault.

13. (B) There are many small tectonic plates and different ways to define a tectonic plate, but the generally agreed-upon number is 15.

14. (A) The "law of horizontality" says that sedimentary rock will always tend to form a horizontal layer regardless of the angle at which lower layers recline. Sedimentary rock will form at a slight angle under special circumstances, but the law of horizontality is generally correct.

15. (D) The hypocenter of an earthquake is the exact spot underground where pressure between two tectonic plates was released.

16. (D) Rayleigh waves are the only side-to-side waves detected on the surface during an earthquake.

17. (E) The Precambrian era makes up about 85% of the earth's history. The Cambrian era makes up the rest.

18. (B) Undersea earthquakes cause ocean floor upheaval that results in tremendous waves.

19. (E) Magnitude 7.0 or greater results in an intensity in nearby settlements of X and greater.

20. (D) The most reliable method by far is the use of historical earthquake records.

21. (A) Younger rock sits on top.

22. (B) Steno discovered that sediments get thinner near the edge of bodies of water.

23. (B) Most divergent plate boundaries are at the bottom of the ocean, although the boundary that runs through Iceland is one exception.

24. (C) A *fault strike* is a line that shows the orientation of a fault with respect to a horizontal plane. It does not account for a fault's inclination from the surface toward the core.

25. (C) A fault zone consists of many related fractures in a single area.

26. (E) Magma generated at a hot spot can form islands.

27. (E) Between 90% and 95% of volcanic gases are water vapor and carbon dioxide.

28. (C) A dormant volcano has not been active recently but may be again.

29. (B) An earthquake's primary effects are its immediate results, such as shaking ground.

30. (C) Searing hot gas is a major hazard of volcanic eruptions.

31. (A) Concerning the law of superposition, sediments take time to accumulate, and they take time to harden into rock under the pressure of higher layers. Each of the higher layers takes additional time to form, as well, and each of these must be younger than buried sedimentary layers. Taken together, these facts support a very old earth—the formation and arrangement of layers could not have taken place in a few thousand years.

According to the law of horizontality, sediment forms horizontal layers that harden into rock. Layers that lie at an angle to the horizontal must be composed of sediment that hardened first and then began to tilt afterward. This extra time-consuming step supports the idea of an ancient earth.

(B) According to Steno's third law, sediments that accumulate in a body of water become thinner at the edges of the body of water. As a result, once buried under later layers, a sediment that had been dispersed across land and sea would thin to nothing at points where the ancient seashore lay. This would appear as breaks in the sedimentary layer.

(C) Taking erosion into account, the Himalayas have risen at an average rate of (8,848 meters) / (10 million years) = 0.88 millimeter per year.

(D) Without uniformitarianism, the Himalayas would have to have risen at (8,848 meters) / (6,000 years) = 1,475 millimeters per year. Yes—the Himalayas would rise by an average of more than 4 millimeters in a single day.

32. (A) Convection currents in the earth's mantle cause the overlying tectonic places to move.

(B) Magma circulates as part of a process called *convection*, in which heat causes deeper magma to expand and rise. As it does this, it cools and begins to fall. The heat is generated by tidal forces on the earth's interior and by the decay of radioactive elements it contains.

(C) (D) These are fracture zones, and at each point where the ocean ridge is offset, there is a transform fracture that is usually perpendicular to the ridge.

33. (A) Because Yellowstone is not near a plate boundary, it's possible to deduce that the area lies over a hot spot.

(B) If previous eruptions occurred to the west and southwest of Yellowstone, then the North American plate is moving to the east and northeast over the hot spot.

(C) The large oval or circular areas are calderas. They form when a magma chamber expels its contents and the ground above it collapses to fill the resulting cavity.

Chapter 2: Atmospheric Conditions

34. (A) Earth's atmosphere is a mixture that is approximately 78% nitrogen, 21% oxygen, and 1% other gases.

35. (C) The Coriolis effect explains differences in the relative direction and speed of wind at different latitudes. Objects such as air molecules at the equator move rapidly relative to objects at higher and lower latitudes, since objects at the equator must move a greater total distance in a single rotation of the earth.

36. (A) The ozone layer within the stratosphere absorbs ultraviolet radiation from the sun. This increases the temperature in the upper stratosphere to temperatures greater than those in the lower stratosphere.

37. (C) The changing seasons on earth are caused by the 23.5° tilt of earth's axis and the movement of earth in its orbit around the sun. They are not caused by the distance between earth and the sun.

38. (B) *Evaporation* is the process by which liquid water changes into a vapor. This occurs as solar energy adds latent heat to water molecules, which allows them to overcome molecular interactions with other water molecules and to escape into the air as vapor.

39. (B) As earth rotates on its axis, objects move at different linear speeds depending upon their latitudes. This is the Coriolis effect.

40. (E) An isobar on a weather map connects areas of equal atmospheric pressure (*iso-* = "equal," and *-bar* = "pressure").

41. (D) Earth's atmosphere is comprised of distinct layers. These layers, in order from the surface of earth into space, are the troposphere, stratosphere, mesosphere, and thermosphere.

42. (E) Precipitation, such as rain or snow, will fall when the humidity in the air is 100%.

43. (E) When people refer to the *relative humidity* on a certain day, they are discussing the connection between the temperature of the air and the amount of water vapor in the air.

44. (D) A *temperature inversion* is a condition in the lower atmosphere where a stable layer of warm air lies above a layer of cooler air. This frequently occurs in valleys and basins.

45. (C) Temperatures on the surface of earth are influenced by many factors. These include elevation, latitude, the distance to a large body of water, and ocean currents.

46. (A) Trade winds are deflected toward and found near the equator.

47. (B) As you ascend to higher altitudes, atmospheric pressure decreases. Therefore, an increase in pressure indicates a drop in altitude.

48. (C) An Atlantic hurricane begins off the coast of Africa with a drop in barometric pressure. This may then build into a tropical depression, which can eventually become a hurricane under the right conditions.

49. (D) The troposphere is the layer of the atmosphere closest to earth's surface. It is this layer that life on earth depends upon. This is where most of our weather occurs.

50. (C) The minimum wind speed needed for a tropical storm to be called a hurricane is 64 knots, or about 119 km/hr.

51. (B) An ENSO event (El Niño-Southern Oscillation) is one in which the temperature of the eastern Pacific Ocean surface experiences a change of more than 0.5°C for longer than five months. These events occur regularly, about every two to seven years, lasting for a period of one to two years.

52. (C) A year in which El Niño is occurring will see warmer than normal water moving eastward to reach the western coast of South America.

53. (A) El Niño and La Niña events are near opposites. Both involve changes in Pacific Ocean surface temperatures, with La Niña involving cool temperatures and El Niño involving warm surface temperatures. Both phenomena change rainfall patterns in parts of the world differently.

54. (E) An ENSO impacts the weather around the globe. ENSO years may see droughts in the western Pacific, floods in lands around the eastern Pacific, and a decrease in Atlantic hurricanes.

55. (A) The eye of a hurricane is the area of calm winds and low barometric pressure in the center of the swirling storm.

56. (D) Tornadoes have clocked the fastest winds of any storm on earth. The average tornado easily reaches wind speeds of 250 km/hr, while the fastest was recorded at more than 480 km/hr.

57. (D) The strength of a tornado is measured using the Fujita Wind Damage Scale. According to this scale, a severe tornado can have speeds of 158 to 206 mph. Storms having slower wind speeds may be called a *gale, moderate tornado*, or *significant tornado*. A faster wind speed would indicate a devastating, incredible, or inconceivable tornado.

58. (A) A jet stream has been compared to a rapidly moving river of air. These are located about 10 to 14 kilometers above the surface. This is within the troposphere.

59. (A) Sunspots are caused by intense magnetic activity in the sun. These bursts of energy most impact the outermost layer of earth's atmosphere, the thermosphere.

60. (B) The ozone layer, residing in the stratosphere, absorbs ultraviolet radiation emitted from the sun.

61. (C) Scientists use many different tools to measure conditions in earth's atmosphere. They use a barometer to measure air pressure.

62. (A) Warm air can hold more water than cold air can.

63. (E) Atlantic hurricanes begin as an area of low pressure off the coast of Africa.

64. (A) The stratospheric ozone layer protects living things from the sun's ultraviolet radiation.

(B) The release of chlorofluorocarbons (CFCs) into the atmosphere causes a problem for molecules of ozone, O_3. Ultraviolet radiation from the sun breaks down CFCs, releasing free chlorine ions that react with O_3 molecules in repeating chain reactions. These reactions can convert O_3 to O_2 and chlorine-oxygen intermediates that continue to react with O_3. This replaces stratospheric O_3 with O_2, which absorbs much less ultraviolet radiation.

(C) The human contribution to ozone in the troposphere comes mostly from burning fossil fuels. In the most common example, nitrogen oxides released by burning fossil fuels create ozone in reactions with volatile organic compounds from automobile exhaust, industrial emissions, chemical solvents, and other sources. Tropospheric ozone can cause many adverse health effects, including drop in crops, lung irritation, asthma, and bronchitis.

65. (A) Earth's seasons are caused by changes in the angle of incoming solar radiation as the earth orbits the sun while tilting on its axis.

(B) During summer in the Northern Hemisphere, the earth is farthest from the sun, but the planet's axis tilts toward it. Sunlight strikes the Northern Hemisphere in the summer at an angle that is closer to perpendicular, resulting in high solar intensity. Although the earth is closest to the sun during the winter, its axis tilts away

from the star, so the sun's rays strike the Northern Hemisphere at a shallow angle that results in less light per unit area.

(C) The climate of a particular spot on earth is dependent upon the weather and the precipitation. These factors vary widely based on the season.

66. (A) Water in the Atlantic Ocean is not warm enough to support hurricane formation until June, and it can stay above the threshold temperature of 25°C until November.

(B) Global warming could result in greater hurricane incidence. According to some models, Atlantic water temperature would rise with average global temperatures, extending hurricane season or increasing the number of hurricanes per year.

(C) When examined over a long period of time, overall hurricane incidence has not changed significantly with changes in climactic conditions.

Chapter 3: Global Water Resources and Use

67. (B) The *Troubled Waters* statement was signed by several marine scientists as a measure to protect marine species from bottom-trawling methods of fishing and other destructive practices.

68. (E) An influent stream is above the water table and is refilled only by precipitation.

69. (C) A *water budget* is *precipitation – evaporation = runoff.* This is a model that includes input, output, and storage of water.

70. (A) The attempts to control flooding on the Kissimmee River led to habitat destruction in the river, the lake into which it flowed, and the Everglades.

71. (D) *Channelization* is a method used to alter streams for uses such as flood control and improved drainage.

72. (B) *Dendritic drainage* is a term used to describe the branching pattern of streams.

73. (C) Since water holds up the soil, when groundwater is removed, the soil sinks, and this process is known as *subsidence.*

74. (A) *Relative humidity* describes the amount of moisture present compared to the amount of moisture the air could possibly hold.

75. (B) Hoover Dam and Glen Canyon Dam are part of the Colorado River.

76. (C) Longshore currents move sand along the beach and barrier islands.

77. (D) Vernal pools are temporary pools of water that are one type of wetlands.

78. (E) Water is held for different time periods in different places. Water is stored for approximately 8 to 10 days in the atmosphere. Other places where it can be stored include oceans, rivers, soil, plants, swamps, and groundwater.

79. (B) Land masses hold 3% of the earth's water, with the remaining 97% in the oceans.

80. (E) Resources such as groundwater that are being consumed faster than they can be resupplied are said to be *nonrenewable*.

81. (B) Ocean salinity increases with depth because cold, salty water sinks.

82. (E) Wind contributes to ocean currents by pushing the ocean's surface. Differences in water density and salinity cause water to sink. Differences in temperature affect ocean circulation because the cooling of warm water also causes it to sink and create currents.

83. (D) The *Gulf Stream* is a current that carries warm Caribbean water north past Canada to Europe.

84. (B) The water table is the upper level of groundwater and is found at the upper edge of the zone of saturation and the bottom edge of the zone of aeration of the soil.

85. (E) Canada and Brazil both have large areas of land and therefore have ample supply of water for their populations.

86. (E) Land that drains to the same lake or river is known as a *watershed*, or *runoff zone*.

87. (A) Preparing a building site produces loose sediment that is washed away in heavy rainfall.

88. (B) Using reclaimed water, preventing runoff, and using dry cooling systems in industrial processes are all methods that help conserve water, also known as *conservation*.

89. (E) When too much groundwater is used up in coastal areas, salt groundwater moves into fresh groundwater, and surface water declines as well.

90. (C) The definition of a drought is rainfall in a given area that is 70% below average for a period of 21 days or more.

91. (D) The definition of an *acre-foot of water* is an acre of water one foot deep.

92. (A) The Watershed Protection and Flood Prevention Act authorized the draining of wetlands to make more farmland and reduce flooding.

93. (B) Dams and reservoirs hurt native fish populations because they flood streams and interrupt migration. They also affect the water quality by lessening the flow of water in a river, causing sediment and salt to build up.

94. (C) The Aral Sea is a body of saltwater that has dried up due to its rivers being diverted for agriculture. Large amounts of water have been withdrawn for irrigation of farmland, killing most of the species of fish.

95. (A) Wetlands have soil that is saturated by water for at least part of the year. The type of wetland area is determined by what type of soil it has and what types of plants grow there.

96. (D) People in the United States use up 40% of the renewable water supply per person per year.

97. (A) One path a drop of water can take is from a rain cloud to falling as precipitation; that rain could then run off the surface of the land to the ocean. The drop would then evaporate from the ocean and eventually form clouds again.

Another path a drop of water could follow is from a rain cloud to falling as precipitation; that rain would then infiltrate the groundwater. The soil would absorb the groundwater. Plants would then take up the groundwater. The water would then transpire from parts of the plant, enter the atmosphere as water vapor, and form clouds.

A third path that a drop of water could take is from rain clouds to rain falling on a lake, with the water then evaporating from the lake and eventually forming clouds.

(B) A stream flows from the mountain and supports fish populations all year. This is known as a *perennial reach* of the stream. Farther on down the mountain, the stream may become ephemeral because it flows in those areas only in response to rainfall.

(C) (i) When too much water is taken out for use by the population, the water table drops. When this happens, trees near the river die, and animal populations die out or migrate because of loss of habitat.

(ii) When trees die off, the banks of a river are much more prone to erosion, and the river shifts its direction, causing damage to man-made structures such as roads. Water floods areas and structures that it was not intended for the river to cross.

(iii) Other changes in the city could include damage to buildings, sinkholes in roads, and settling of the land.

98. (A) An example of a natural event that contributes to flooding includes a situation in which the land has little vegetation, causing the precipitation to run off the surface, as it does in flash floods in the desert. Man-made changes that contribute to flooding include deforestation from logging and also paving of surfaces in urban areas, both of which increase surface runoff.

(B) Two methods of flood control that are currently in use in the United States include levees and streambed channelization. Levees hold water back from flooding an area. This is a problem because if they fail, there is nothing else to prevent destruction. Streambed channelization occurs when the sides of a river are bulldozed to make them wider and allow more water to flow. This damages habitats, causes erosion, and creates flooding somewhere else.

(C) The Watershed Protection and Flood Prevention Act of 1954 approached the flooding issue with the idea of draining wetlands. This took care of the problem in the immediate area but then caused flooding in other places.

(D) To implement watershed management, there are several methods that can be employed. Development should be limited, and areas that have been deforested should be replanted. Also, surface runoff should be diverted to holding areas.

99. (A) Dams are useful to people, despite the environmental destruction that ensues. First, dams store water for future use. This is especially important to ensure an adequate water supply for people in regions with little precipitation and also as the population continues to grow. Second, dams provide hydroelectric power to cities, offsetting the need for fossil fuels and therefore reducing carbon emissions. Third, dams provide recreation areas for people for activities such as fishing, boating, and waterskiing. These areas bring in tourist revenue for local towns and also provide jobs.

(B) Dear residents of Springfield,

As you know, voters will be deciding the fate of the new dam in the upcoming election. I am writing to you today to urge you to vote "no" on this measure for several reasons. We can learn from the example of the Three Gorges Dam in China what impact the proposed dam could have on the future of Springfield. First, flooding is a major issue. People in China built settlements downstream from the dam despite the fact that they were in a potential flood-prone area. This is especially problematic since the area is susceptible to earthquakes and has already experienced major landslides. In addition, the sedimentation of the river is destroying the habitat of endangered species and contributing to the previously mentioned risk of flooding. The river that supplies the dam is already being polluted with sewage, which stagnates in the dam. Hundreds of thousands of residents were relocated to build Three Gorges Dam and were never compensated. Archaeological sites became flooded, and the beauty of the land was forever altered. Is this the future we want for the town of Springfield? Please vote "no" on construction of the Springfield dam.

(C) Although there are some advantages to having a dam, there are many ways to manage the water supply to avoid the need for new dams. First, the number of people in arid regions must be limited if possible. This includes not developing land for housing in areas that don't have an adequate water supply. Second is water conservation. Irrigation systems in agricultural areas need to be updated to use the most efficient and conservative methods. Crops that require a lot of water should not be grown in dry regions. Industrial water use should be updated to be efficient as well. Individuals should change their habits in their own homes, including watering landscape at the right time of day and in moderation and installing low-water-use showerheads, toilets, and washing machines. Third, cities should recycle water and irrigate the public landscape with recycled wastewater. And last, governments can assist by offering incentives for water conservation and passing laws that mandate watering restrictions and higher rates for excessive water use.

Chapter 4: Soil and Soil Dynamics

100. (B) Sedimentary rocks are deposited, or laid down, in layers. Igneous rocks form from cooling magma or lava, and metamorphic rocks form from the addition of heat and pressure to existing rocks.

101. (D) Diagenesis, the process by which sediments are changed into sedimentary rock, has several steps. These steps include compaction, cementation, recrystallization, and chemical changes. When a rock is melted, it becomes magma or lava, which will harden into an igneous rock.

102. (A) Igneous rocks are classified according to their texture and composition. Igneous rocks may be mafic or felsic in composition.

103. (B) When any rock melts, due to burial, mountain-building pressures, or other heat sources, the rock becomes magma.

104. (B) Physical weathering is the process by which a rock (or something else) is broken into smaller pieces without changing the chemical composition.

105. (D) Chemical weathering breaks substances down by changing their chemical composition. The process of hydrolysis chemically weathers through a reaction with water.

106. (C) Some minerals dissolve in water. Solubility, a form of chemical weathering, is the ability of minerals to dissolve in water.

107. (E) The uppermost layer in a soil profile, the A horizon, contains the most organic material. This area is very rich in nutrients and can support many plants and small animals or insects.

108. (E) Plants can break down rock by physical weathering, as roots push through rock, breaking it apart. Animals, such as worms or moles, may push through rock and soil, physically weathering as they go. Many plants and animals may secrete chemicals that will slowly chemically weather a rock.

109. (C) A pedalfer soil is located in the B horizon. These are clay soils rich in iron oxide and aluminum.

110. (D) Laterite soils, characterized by signification leaching and an accumulation of calcite and organic materials, develop in hot, tropical climates.

111. (E) The bedrock is the parent material of a soil profile. Often, large recognizable chunks of bedrock can be found partially weathered in the C horizon.

112. (D) The uppermost two layers of a soil profile make up the topsoil. These are the O horizon and the A horizon.

113. (A) The B horizon is beneath the A horizon. The A horizon is also referred to as the *zone of leaching* because material such as soluble minerals are moved through the layer with water that passes through. The leached material ends up in the B horizon, making it the zone of accumulation.

114. (B) Pedocal soils are found in dry areas such as the western United States. These soils are characterized by significant accumulations of calcium carbonate.

115. (A) *Contour plowing* is a technique that is used to prevent soil erosion. The other choices are examples of situations that can increase erosion.

116. (D) *Terracing* is the practice of taking one very large, very steep field and breaking it into many smaller fields. This helps prevent large amounts of soil from eroding from the surface.

117. (C) Frost wedging occurs when water falls into a crack in the rock and then freezes as temperatures drop. When the temperatures rise, more water enters into the now-larger crack, and when it refreezes, the crack is expanded further. This is a type of mechanical weathering.

118. (E) Salt wedging is an important type of mechanical weathering of rock in a desert.

119. (A) Dolomitization is what occurs when sedimentary minerals dissolve and are changed into other compounds. Typically, limestone is changed into dolomite during this process.

120. (C) Clastic sedimentary rocks, such as sandstone or conglomerate, are composed of small fragments of other rocks or sediment.

121. (B) *Lithology* is a visual investigation of the physical characteristics of a rock. These characteristics are studied using a hand lens or a microscope.

122. (E) When unloading occurs, the pressure is released on a rock inside the crust, as the rock layers above it are worn away by erosion and mass wasting.

123. (C) Plants and animals are known to weather rock by both physical and chemical means. They may physically break down rock by forcing their way into a small crack, and they may release chemicals that will help to break down the rock.

124. (A) Frost wedging is the process by which water enters into a crack in a rock, freezes, and then expands, causing the crack to get larger. The water eventually melts and then refreezes, increasing the size of the crack each time it does. This process is most prevalent in an environment that is cool and wet and has a cycle of freezing and thawing.

125. (B) When rainwater mixes with carbon dioxide in the atmosphere, carbonic acid is formed.

126. (B) Contour plowing is the process of planting crops along equal elevations on a hill rather than up and down the hill. This slows down the runoff of water.

127. (B) The most common kind of oxidation of rock is the formation of rust in rocks that contain iron.

128. (D) Breccia is a sedimentary rock formed when pebble-sized particles that are not rounded due to erosion are compacted and cemented together.

129. (D) Granite, when it undergoes metamorphosis, becomes gneiss.

130. (A) Conventional agriculture regularly uses synthetic herbicides, pesticides, and fertilizers. Sustainable agriculture avoids these substances and instead substitutes solutions that have a smaller impact on the environment. These solutions include removing weeds by hand or using natural herbicides, controlling insect pests with insect predators, and using compost in place of chemical fertilizers.

(B) Conventional agriculture can spoil the land by destroying habitats as the area is cleared for agriculture; using fertilizer, pesticides, and herbicides, which can contaminate the groundwater; and exposing the soil to further erosion.

(C) Sustainable agriculture avoids disruption of the land by maintaining natural ecosystems; by not clearing entire areas, and thus not leaving the land bare; by planting diverse crops, using natural fertilizers, and preventing chemical pollution of land and water; and by consuming less energy for chemical fertilizers and other synthetic substances.

131. (A) Physical weathering processes include frost wedging (water infiltrating a crack in a rock, expanding as it freezes, and enlarging the crack), salt wedging (salt crystals forming as water evaporates, expanding, and putting pressure on a rock), and unloading (rocks that have been buried under pressure coming to the surface, expanding, and cracking). Chemical weathering processes include oxidation (the reaction of a substance with oxygen), hydrolysis (the reaction of a substance with water), and acid precipitation (the reaction of a substance with acidic rain or snow).

(B) Weathered sediment can become sedimentary rock if it is buried and compacted. Often a natural cement, such as calcium carbonate, hematite, or quartz, acts to further bind the particles together into a sedimentary rock such as sandstone.

(C) Sediment can be compacted or cemented to form sedimentary rock. Sedimentary rock can be weathered back into loose sediments again, or it can go on to become igneous or metamorphic rock through the rock cycle. Sedimentary rock that melts into magma hardens into an igneous rock as it cools. Sedimentary rock that is subjected to heat or pressure can turn into a metamorphic rock.

132. (A) Igneous rocks may be *intrusive* or *extrusive.* These terms describe where the rock forms and therefore how large the mineral crystals are. Intrusive igneous rocks form as magma slowly cools beneath earth's surface over time. The crystals in an intrusive igneous rock tend to be large enough to see without the use of a microscope or hand lens. Extrusive igneous rocks form from lava that has erupted onto earth's surface. The lava cools quite quickly, which prevents large crystals from forming. These crystals are typically not visible to the unaided eye. Igneous rocks are also classified according to their composition. Mafic rocks are dark colored, with minerals such as amphibole and biotite mica. Felsic rocks are light colored, predominantly composed of minerals such as quartz and feldspar. Granite is a light-colored, intrusive igneous rock. Basalt is a mafic, dark-colored, extrusive igneous rock.

(B) Sedimentary rocks form as sediment is buried and slowly compacted over time. Many times, some sort of cementing agent helps to bind the sediment together. Common cements found in sedimentary rock include calcium carbonate, quartz, and hematite. Sandstone is a sedimentary rock formed as sand-sized grains of sediment are compacted and cemented. Shale forms in quiet environments such as a lake or deep ocean, as silt- and clay-sized sediment slowly settles out of the water, compacting into thin, friable sheets.

(C) Metamorphic rocks are other rocks that have been changed due to the action of heat or pressure. One important thing to note about metamorphic rocks is that the heat or pressure cannot be so intense that the rock melts; if this occurs, the rock will then become an igneous rock, not a metamorphic rock. Some metamorphic rocks are derived from other metamorphic rocks. Others may form as igneous rocks, such as granite, are subjected to the heat and pressure of mountain-building processes. Granite, as it is metamorphosed, becomes gneiss. Gneiss is a banded rock, resulting from granite's heterogeneous composition. The sedimentary rock limestone becomes marble, a popular building stone, as it is subjected to heat and pressure. Marble consists primarily of one mineral, calcite, so it has a mainly homogeneous composition.

Chapter 5: Ecosystem Structure and Diversity

133. (B) The biosphere encompasses living and nonliving components, while ecosystems include the connections between organisms in a large area, and communities include all the organisms in a particular habitat.

134. (C) A greater diversity of genes confers a greater diversity of form and behavior.

135. (A) Only the upper crust is counted as part of the biosphere, and it is closer to the core than any other part of the biosphere.

136. (D) Genetic differences account for most variation in tolerance to the physical and chemical environment. Acclimation can increase an individual's tolerance but only within the limit set by genes.

137. (D) Each of these is a measure of diversity among organisms.

138. (E) The other four components are essential to any ecosystem.

139. (C) Each organism generally obtains only 10% of the energy contained in a consumed species.

140. (E) The highest predators are the most dependent upon a stable ecosystem.

141. (A) Along with tropical rain forests, swamps and estuaries have the highest net primary productivity of all ecosystems by far.

142. (B) Physical damage does not affect the niche where its species survives and interacts with its ecosystem.

143. (C) Biologists gauge fitness in terms of offspring.

144. (B) A bottleneck event has likely damaged genetic diversity, reducing the population mostly to green lizards.

145. (A) Beneficial mutations are rare, harmful mutations are more common, and harmless mutations are more common still.

146. (D) Evolution is generally identified through a change in allele frequencies in a population over time.

147. (D) Speciation is the formation of a new species from a splinter group, generally including a loss in the ability to reproduce with the larger original group.

148. (A) In evolutionary divergence, one species splits into several species that limit competition with each other.

149. (C) Buffalo were once endemic to the North American West, while dodo birds were once endemic to Australia.

150. (E) A habitat includes only the tangible elements of an organism's environments.

151. (C) Creatures with narrow niches are less able to switch to a new food source or a new type of shelter, for example.

152. (A) Commensalism is a relationship in which one partner benefits to neither the loss nor gain of the other partner.

153. (D) Trees are latecomers in colonizing an area, arriving only later in secondary succession. There is no tertiary succession.

154. (A) Lichens can live on bare rock with very little water and slowly extract nutrients from the rock, air, and passing debris.

155. (C) Removing keystone species can disrupt big portions of the rest of an ecosystem.

156. (E) Large marine mammals do not generally live in coastal wetlands but in the open ocean or in coastal areas that do not have enough vegetation to be considered a wetland.

157. (B) The amount of sunlight that each zone receives is generally a function of its depth, with the abyssal zone getting the least light, and the euphotic getting the most.

158. (D) The littoral zone includes submerged soil near the surface along the edges and shallow areas of lakes.

159. (C) Hot spots are a key part of the emergency plan to save as much terrestrial diversity as possible. Together they account for about two-thirds of such diversity.

160. **(E)** Restoration always involves attempting to return an area to its previous, unspoiled state—while creating a wetland from a ruined grassland can be a good idea, it's not restoration.

161. **(B)** Costa Rica is the world leader in protecting the greatest proportion of its biodiversity.

162. **(A)** Phytoplankton float in the sun-drenched euphotic layer, where they can photosynthesize most efficiently.

163. **(A)** The silversword alliance is an excellent example of the founder effect, which is a variation of the bottleneck effect. Each species in this set is descended from a common ancestor that arrived in Hawaii very recently, in geological time. Each species has had time to acquire only minor genetic differences from that first ancestor and still shares most of the same genes in nearly the same form as they arrived in the Hawaiian archipelago.

(B) Early silversword alliance ancestors began to colonize larger areas of the Hawaiian archipelago, and those that managed to occupy a new niche, such as in the low-nutrient soil of Mauna Kea, often experienced less competition than their relatives living in more crowded niches. Gradually these plants became specialist species as their traits evolved to better fit the new niches.

(C) *Argyroxiphium sandwicense* is involved in primary succession because it inhabits nearly lifeless areas covered by volcanic rock.

(D) Because the Hawaiian Islands are difficult to reach, the herbivorous mammals were almost certainly brought by people coming to the island recently, so the plant has not had to deal with this kind of pressure on its population until recently. It takes a longer time than a few hundred years for protective genetic adaptations, such as thorns, to evolve.

164. **(A)** The common grackle probably left Central Texas due to habitat loss. The introduction of wider spaces and different vegetation favored the great-tailed grackle, and the great-tailed bird's larger size probably gave it a competitive advantage over its smaller cousin.

(B) The boat-tailed grackle's realized niche is every part of its coastal wetland niche in which it does not compete with the great-tailed grackle or other species for space, food, shelter, water, and other necessities. The boat-tailed grackle's fundamental niche includes these areas, as well as those in which it competes with the great-tailed bird.

(C) The boat-tailed grackle is likely more specialized to living along coastal areas than the great-tailed grackle, and it probably experiences less competition from the great-tailed cousin than the common grackle did, once great-tails moved in.

165. (A) *Mutualism* is the best name for such a relationship because both partners gain and neither appears to suffer as a result of the association.

(B) In areas with only bare rock—or in fact in any area that lacks other living things—lichens are pioneer species. They slowly adapt an area to conditions that other species find habitable. For example, some lichens living on bare rock will slowly dissolve the rock into soil.

(C) A reindeer eating lichen has a parasitic relationship with the lichen—the lichen gains nothing and loses a lot, while the reindeer gains without any known negative consequences.

Chapter 6: Natural Cycles and Energy Flow

166. (C) The term describes the cycling of several different elements through the earth's organisms, atmosphere, soil, and rock. Biologists are generally most concerned with carbon, sulfur, nitrogen, water, and phosphorus.

167. (C) Only high-energy physics phenomena, such as atomic fusion, can convert matter to energy or energy to matter.

168. (B) Nitrogen is very chemically unreactive, so the average N_2 molecule remains unchanged for a long time compared to compounds made with other biologically important elements.

169. (E) Carbon forms the "backbone" of such biologically important molecules as DNA, proteins, lipids, and carbohydrates. Also, the term *organic chemistry* refers specifically to the chemistry of carbon-based compounds.

170. (A) Marine organisms' shells collect on the ocean floor, eventually creating a layer of calcium-carbonate sedimentary rock. When some of this rock is carried into the earth's interior at a subduction zone, it releases carbon dioxide as a result of high heat and pressure—the calcium typically does not change to a gaseous form. Volcanoes can release large amounts of carbon dioxide from deep in the earth.

171. (B) Very little carbon has floated off into space—the vast majority has continued to cycle through earth's geochemical cycles since the planet was young.

172. (D) Like all organisms, plants break down carbohydrates and sugars for energy during respiration. Their photosynthesis creates these molecules using sunlight, and along with all other constituent molecules, these carbon-containing molecules break down into CO_2 and other products when they decompose.

173. (E) Plants use sunlight to make sugars and carbohydrates.

174. (B) Phytoplankton rely on carbon dioxide dissolved in the water. It becomes dissolved in water due to the water's contact with the air.

175. (D) The composition of the earth's core is metallic, primarily iron and nickel, and it does not generally exchange material with the mantle.

176. (D) The geological carbon cycle needs shell-forming organisms to create the layers of limestone that can be pushed into the earth's mantle during subduction.

177. (A) Photosynthesis takes carbon dioxide from the air to make sugars and carbohydrates, while respiration does the reverse. Due to the enormous amount of carbon that plants, animals, and other organisms use every year, any imbalance of these two phenomena can cause pronounced differences in atmospheric carbon dioxide year to year.

178. (D) At deeper parts of the ocean, higher pressure allows more carbon dioxide to dissolve in the water. Dissolved carbon dioxide makes water more acidic, and these deep parts of the ocean are just acidic enough to dissolve calcium-carbonate shells.

179. (A) Limestone and other types of rock contain more calcium than any other natural source, and calcium does not generally form large amounts of any gas spewed by volcanoes. Phytoplankton obtain calcium from the water.

180. (A) Nitrogen makes up about 78% of the atmosphere, by volume.

181. (E) Only high-energy electrical discharges and nitrogen-fixing bacteria naturally create biologically usable nitrogen compounds.

182. (D) Destroying vegetation keeps an area from absorbing carbon dioxide through photosynthesis, while decaying plant matter releases more of the gas into the environment. At the same time, destruction of a vegetation-rich ecosystem releases nitrogen compounds all at once into the environment, while these molecules would ordinarily remain locked in organisms or be rapidly reabsorbed when released by small numbers of decaying organisms.

183. (C) Certain bacteria convert NH_3 to NO_2^- and NO_3^-. Plants can use NH_3 and NO_3^- but not NO_2^-.

184. (A) Nitric oxide, NO, from burning fossil fuels is converted to NO_2 and HNO_3—nitric acid—in the atmosphere, which can contribute to ocean acidification.

185. (D) Carnivores must get their nitrogen from consumed animals. Nitrogen compounds enter animals when primary consumers—herbivores—eat plants.

186. (B) Rock is the major source of phosphate on land.

187. (D) Low nutrient levels, particularly of phosphate, keep algae populations in check. When more than enough nutrients are available, algae grow quickly and use up dissolved oxygen, a process that can suffocate many marine organisms.

188. (A) Fossil fuels contain all these types of pollution.

189. (C) Whether in clouds, precipitation, runoff, or lakes and oceans, water remains H_2O throughout the water cycle.

190. (C) Sunlight causes nearly all evaporation.

191. (B) *Percolation* describes the flow of water to underground storage.

192. (C) *Transpiration* is evaporation from plants that have absorbed water, usually through their roots.

193. (C) In order for cloud droplets to form, water vapor needs a particle to collect on, such as dust.

194. (E) After evaporating from the ocean, some of this water is transported in the form of clouds that release precipitation over land.

195. (C) Oceans comprise about 75% of the earth's surface, but being water, they release more evaporation by area than land does. As a result, about 85% of evaporated water comes from oceans and other water bodies.

196. (B) Ocean sediments and sedimentary rocks hold more than 1,000 times as much carbon as any other source.

197. (C) The earth's oceans contain up to 1,000 times as much calcium as the soil, while the atmosphere contains negligible amounts.

198. (B) Decomposer bacteria break complex molecules, such as proteins, into smaller molecules and ultimately into nitrate and nitrite ions.

199. (A) Smelting ores and burning fossil fuels release sulfur.

200. **(A)** Both tertiary consumers and secondary consumers eat other consumers, while primary consumers are herbivores and eat only plants.

201. **(A)** Carbon is incorporated into the shells of marine organisms in the form of calcium carbonate. The shells can become part of seafloor sediment that is carried down into a subduction zone by a descending oceanic plate. When the plate reaches a depth at which melting begins to occur, the resulting magma can rise through the overlying plate and erupt through volcanoes. Some of the carbon in this magma comes from marine sediments.

(B) Sulfur dioxide becomes oxidized to sulfur trioxide, and the addition of water changes it into sulfuric acid, H_2SO_4. This can reach land through acid deposition—acid rain—where sulfuric acid can be neutralized to become a sulfate salt, such as Na_2SO_4. Plants take up these salts and use their sulfur as a component of proteins.

(C) Carbon dioxide from volcanoes can enter organisms in the same way that other carbon dioxide does—it's incorporated into carbohydrates by plants and other photosynthetic organisms. Plants—or consumers—metabolize these carbohydrates for food and either incorporate the constituent carbon into other biological molecules, such as DNA, or release it as the waste product carbon dioxide.

202. **(A)** Nitrogen-containing organic matter, such as dead plant and animal material and topsoil, is often washed out to sea. Even if bacteria can convert organic material in the ground and nitrogen gas in the air into nitrates that plants can use, much of this is lost in runoff.

(B) With little of the vegetation it once had, Madagascar doesn't contribute as much water vapor to the hydrologic cycle as it once did through transpiration and evaporation from land. Instead, a great deal of water runs off the land nearly unimpeded into rivers and out to sea.

(C) With less usable nitrogen and captured water in the soil, plants are not able to grow as quickly, so Madagascar's plants as a whole capture less carbon dioxide as they photosynthesize carbohydrates. At the same time, a great deal of carbon-containing organic material is lost in runoff, so the island doesn't contribute as much carbon dioxide to the carbon cycle from organic matter decaying on land. Instead, this ends up as carbon dioxide that joins the biological carbon cycle once it is broken down in the sea.

203. **(A)** The cave ecosystem ultimately depends upon producers in the outside world—plants that photosynthesize using sunlight and are eaten by consumers (or a series of consumers) that die or leave droppings in caves. However, the lowest level of the cave food chain is the decomposer fungi and bacteria, which serve an underground role that is similar to the role of producers.

(B) Primary consumers in the caves are those creatures that eat bacteria and fungi, such as millipedes and flatworms.

(C) The secondary consumers in this example are spiders and salamanders.

Chapter 7: Population Biology and Dynamics

204. (B) A population is a large group of one species, while an ecosystem includes all interacting species.

205. (D) *Fecundity* refers to a species' reproductive potential—its ability. *Fertility* refers to the actual number of offspring an individual or population has produced.

206. (D) *Life expectancy* is a statistical measure often used to determine future demographics by estimating how long individuals in a population will live.

207. (D) Clumping is the most common form of population distribution.

208. (C) Only births and immigrations add new individuals—*emigration* refers to individuals leaving a population.

209. (A) By definition, the environment cannot carry more individuals of a given species than its carrying capacity, so the population must come down.

210. (D) Logistic growth tends to keep populations varying slightly above and below the carrying capacity.

211. (C) At and around point V, the curve's slope continues to increase as time passes.

212. (B) Population 1 enters logistic growth centered around the level of 10,000 individuals, which must be the carrying capacity.

213. (E) Population 1 varies from slightly above to slightly below 10,000 individuals without making major changes in either direction.

214. (A) Due to the later crash of Population 2, we can infer that its peak near point Z is above the environment's carrying capacity.

215. (D) Logistic growth tends to continue to approximate the environment's carrying capacity.

216. (D) Earthquakes affect individuals whether or not they are densely distributed. The other events occur more often in dense populations.

217. (B) Populations can fluctuate slightly and still be considered stable.

218. (D) At the industrial level of development, human societies often have low birthrates, due to education of women and other factors, and low death rates, due to more advanced health care, safety standards, and other factors.

219. (A) An r-selected species relies more on producing a quantity of smaller offspring that require less energy per individual to produce rather than single, large, long-lived offspring that require a relatively greater amount of energy per individual.

220. (C) Genetic drift is essentially random, producing genetic diversity in greater or lesser amounts, whereas the other listed phenomena result only in less diversity.

221. (E) When two populations join, they can trade genes widely, dispersing more traits among the whole enlarged group.

222. (B) Two populations that trade members, and thus are genetically connected, are called a single *metapopulation*.

223. (D) Most developed nations besides the United States are projected to produce fewer babies than they already have older individuals, who have a higher death rate. That is, the large older population is expected to contribute to a higher death rate for these countries, and negative growth is likely to result from birthrates that can't match death rates. This can lead to resource allocation problems, such as a decreased tax base from which to draw social security funds.

224. (D) K-selected species tend to colonize areas in late succession, have large offspring, and occupy a specialist niche in the ecosystem.

225. (C) Several variations exist, but the three general survivorship curves are constant loss, early loss, and late loss curves. These reflect the ages at which individuals of a species tend to die.

226. (C) The most likely reason that Country X has larger age brackets at the younger end of its population is that the country has a higher birthrate than Country Y.

227. (A) With a disproportionately large segment of its population approaching retirement age, Country Y is likely to lose much of its tax base as older workers stop working and young workers do not replace them in comparable numbers.

228. (B) With a large segment of its population just below reproductive age and industrialization lowering its overall death rate, Country X is likely to grow more quickly in 15 years.

229. (C) With minor variations between age brackets, a constant loss survivorship curve would result in a population of decreasing numbers in older age brackets.

230. (B) With few exceptions, societies with slowing or stagnating growth rates are postindustrial.

231. (C) Lower birthrate results in fewer people in low-end age brackets. Emigration would result in mostly older people leaving the population, while immigration—with few associated births—would have to occur at a very high rate to result in such a lopsided age structure.

232. (B) Fecundity is an organism's reproductive capability, while fertility is the actual number of offspring it has produced.

233. (C) (100 + 20 – 10 + 30) mockingbirds / 5 square kilometers = 28 mockingbirds per square kilometer.

234. (A) Wild pigs conform more to the characteristics of r-selected species. They produce many small offspring very quickly and have an early reproductive age—and thus have a high population growth rate. They often die before reaching reproductive age and are highly adaptable generalists. However, wild pigs don't have all the characteristics of r-selected species—for example, they show parental care for their offspring, and they grow to become large adults.

(B) The early loss survivorship curve best describes the mortality of wild pigs. After the pigs mature to adulthood, they tend to survive.

(C) No. Wild pigs' populations were still growing exponentially well into the 2000s. Approaching the environmental carrying capacity results in a population decrease soon afterward.

(D) The predators and hunters mentioned in the paragraph comprise environmental resistance to the wild pig population.

235. (A) During that period, the U.S. Northeast had a net emigration of 89,000 people. That is, the population shrank by 89,000 people.

(B) The West had greater immigration by 630,000 people. That is, 630,000 more people moved to the West than to the Northeast between 2003 and 2004. The Northeast had greater emigration than the West, by 119,000 people.

(C) The fertility rate of women during this period was 3.7 million births per 148.9 million women, or about 2.5%.

(D) The fecundity rate during this period among women was at least 61.6 million births per 148.9 million women, or at least 41.4%.

236. (A) The Great Plague was a density-dependent factor affecting certain people in England—those who lived closely together, especially in London. Diseases are less able to spread among widely spaced populations.

(B) The Little Ice Age is a density-independent factor, since it affects a broad area of humans without regard to how close they live to one another.

(C) The plague is the only biotic factor, while the Little Ice Age and the Great Fire of London are both abiotic factors.

Chapter 8: Agriculture and Aquaculture

237. **(C)** A lack of vitamin A causes damage to the eyes and can lead to blindness.

238. **(A)** Marasmus results from a deficiency of both protein and calories, whereas kwashiorkor results from just a protein deficiency. The most visible difference is that people with marasmus are active and those with kwashiorkor are passive in activity. Marasmus victims also have ribs protruding, whereas kwashiorkor victims have protruding bellies.

239. **(B)** Obesity is defined as being more than 30 pounds overweight for a person's height and age and affects about one-third of the population. Often this is the result of eating too much protein, salt, sugar, and fat but not enough fiber.

240. **(A)** About 82% of soybeans, 71% of cotton, and 25% of corn are GMOs.

241. **(D)** Drip irrigation is the most efficient way to provide water directly to a plant's roots and reduce runoff and waste.

242. **(D)** Microbial/biological agents used as pesticides include Bt, ladybugs, lacewings, and wasps.

243. **(E)** Strip cropping is a farming method in which two different crops are alternated in strips on the same plot of land so that one crop can slow the runoff of water from the other and also act as a wind barrier.

244. **(C)** Subsistence crops are grown by the farmer primarily to directly meet the farmer's own food needs.

245. **(E)** Perennial crops, which grow from the same root system each year, have many ecological benefits despite having slightly lower productivity.

246. **(B)** Europe has about 30% of its land in use for agriculture.

247. **(B)** The National Germplasm System holds seeds for study and possible use.

248. **(D)** Both the Food and Drug Administration and the Department of Agriculture enforce the pesticide limits of the Environmental Protection Agency.

249. (A) Atlantic salmon are endangered from being farmed because of water pollution from their waste products and food, as well as from interspecies breeding.

250. (C) A limiting factor is the one nutrient a crop needs that is in the shortest supply.

251. (E) Revegetation of land in China has helped decrease wind erosion and reverse local climate change.

252. (B) Topsoil is beneath the litter layer, has both organic and inorganic materials, holds moisture well, and supports crop growth.

253. (D) By rotating different crops and allowing the land to "rest" or be left fallow from harvest occasionally, farmers reduce the prevalence of disease and prevent the soil from getting depleted of nutrients.

254. (B) An increase in carbon dioxide levels causes a decrease in nitrogen levels in all plants, reducing the overall quality of the plant for the rest of the food chain.

255. (A) Improving transport, refrigeration, storage, and pest control of crops would help offset the losses of almost one-third of agricultural crops.

256. (C) Approximately 60% of the pesticides found in our waters come from herbicides, the largest share of any of the pesticides.

257. (B) *Silent Spring* was written by marine biologist Rachel Carson. It pointed out the environmental contamination caused by careless use of pesticides and herbicides.

258. (A) The Clean Water Act was enacted in 1977 to address concerns about water pollution.

259. (E) Approximately 10,000 years ago, humans began agriculture and animal domestication. This was a turning point because it started the significant alteration of the environment.

260. (D) Biological diversity is decreased due to loss of competing ecosystems, not increased.

261. (B) Secondary succession takes place when an ecosystem has been removed. Slash-and-burn agriculture creates this environment, and the diagram shows the growth that takes place following this event.

262. (C) People in China are consuming more meat than before, which requires more grain to be produced to feed the livestock.

263. (C) Minimum tillage is a farming technique that reduces the manipulation of the soil. This significantly reduces soil erosion because the soil is held in place by crops or residues of crops.

264. (A) Planting crops such as alfalfa and clover in the off-season and then plowing them under provides an organic fertilizer with many benefits that are missing in synthetic fertilizers.

265. (D) Pasture is land that is plowed and planted for livestock. Rangeland is used for feeding livestock as well but is not plowed or planted.

266. (E) Raising livestock requires water to grow grain and then feed the grain to the livestock. This consumes more than one hundred times the water needed to just grow most grains alone.

267. (A) Three signs of desertification are reduced water in ponds and streams, increased erosion, and lowering of the water table.

(B) Two specific examples of people's contribution to desertification include too much farming in an area and deforestation of an area for timber.

(C) Instead of overfarming, people could practice crop rotation and leave some areas fallow so their fertility can be restored. Also, instead of deforesting an area for timber, local people could utilize the forests for other products such as coffee, medicine, and native crops, leaving the forest intact. They could meet their needs by trading with each other or by partnering with Western countries to provide these goods for sale outside the country.

268. (A) Bottom trawling is a method in which a net is dragged along the bottom of the ocean to catch fish, but it often catches many unintended species. Discarding bycatch is used when fish that are caught but are the wrong type or size are thrown back overboard; but often they are dead before they can be returned.

(B) The development of better fishing gear that doesn't trap so many unwanted species from the bottom is one method of preventing overfishing. Also, raising fish on farms may prevent breeding stocks from getting depleted.

269. (A) There are several problems Farmer Jill could have with choosing the first property. Two of these problems include ridding the land of pesticides and herbicides that are already in the soil and converting mechanized irrigation to drip irrigation. Additional problems include replenishing the soil, which would be lacking in nutrients from use of industrial fertilizers and overuse, and rehabilitating the soil structure that would be damaged from tilling.

(B) The second property would have used contour plowing methods. It would have drip irrigation, and the soil would not have been tilled. The previous farmer would use natural enemies of pests such as ladybugs and lacewings. The use of pesticides would be limited.

(C) (i) Creating the new organic farm would mean using no pesticides, using no artificial fertilizers, and no tilling of soil. Drip irrigation and alternating and interplanting crops would also be important.

(ii) Despite the many benefits, the drawbacks of owning an organic farm include cost of organic certification, cost of organic chemicals, reduced yields due to rotating or resting the soil, and the loss of crops due to insects and weeds.

(iii) The principles of integrated pest management include control but not eradication of pests, the use of natural agents, and the concept of managing an ecosystem instead of dealing with each species separately. The methods of integrated pest management include use of only natural, Organic Materials Review Institute (OMRI)–certified chemicals, more types of crops planted together, use of natural enemies of pests (ladybugs, lacewings, décolleté snails), no-till agriculture, and development of genetically resistant plants.

Chapter 9: Forestry and Rangelands

270. (B) Temperature and climate determine if a forest is a temperate rain forest, a tropical dry forest, or a tropical rain forest.

271. (A) A temperate rain forest has conifer trees, large amounts of rainfall, and cool temperatures. This is found only in a few small regions of the world.

272. (D) Approximately one-half of the original rain forests have been cleared, with loss continuing at a rate of 1.8% per year.

273. (C) *Selective cutting* is deliberately thinning mature trees in order to leave a seed source for new tree growth and provide habitat as well.

274. (E) Grasslands are often converted to cropland or urban areas and often at a rate of three times that of tropical rain forests.

275. (A) Steppes are not as dry as deserts and contain grassland plains.

276. (C) Forest fires are beneficial in opening cones to release seeds, removing brush, replenishing soil, and *preventing* crown fires from occurring, not causing them.

277. (B) Drought is a natural event. Degradation occurs only with unsound human practices.

278. (D) Modern environmentalism looks at the picture of the whole ecosystem in order to preserve a particular area or species.

279. (A) Nitrogen, in the form of nitrates or ammonia, produces proteins that are used to make enzymes and other compounds to help the plant live.

280. (B) The Public Rangelands Improvement Act of 1978 improves the range conditions and prevents overgrazing.

281. (B) Sustainable forestry practices involve cutting only enough trees so that the ecosystem recovers and the following year it is possible to cut the same amount of trees.

282. (C) The U.S. Forest Service was established by President Theodore Roosevelt in 1905.

283. (E) Wilderness area is land that is mostly not affected by humans.

284. (D) Untouched areas of forest were designated as wilderness areas, and human activities in the forest were regulated.

285. (A) Indirect deforestation occurs when trees die from causes other than logging. Pollution, disease, and global warming are possible causes of indirect deforestation.

286. (E) The metabolic reserve is the minimum portion of the plant that must be left after grazing by livestock so that the plant can stay alive.

287. (C) Suppressed trees are those that are not reached by direct sunlight because they are lower than the other layers of trees in the forest.

288. (D) Public service functions of forests are benefits that mankind receives from the forest just by the forest's existing. Harvesting timber requires work and is not simply a by-product of the forest's existence.

289. (A) Only 0.1% of tropical forest logging uses sustainable practices. Much of the industry involves clear-cutting forests and some selective cutting.

290. (C) Water content of soil and need for more or less sunlight are two types of factors that determine a tree's niche.

291. (B) The soil of grasslands has a lot of organic content from decaying plant material.

292. (E) Horses, cows, and other domestic animals were brought to North America by European settlers and often brought disease and weeds with them.

293. (D) Feedlots are places where cattle are kept in large quantities in small spaces. Their waste often pollutes local water supplies.

294. (B) Developing countries use most of their wood for firewood as opposed to developed countries, which use it mostly for construction, paper, and furniture.

295. (C) Parks are designed for people to enjoy and use. Although they often have several other functions, such as preservation, in mind, the area is set aside for people.

296. (A) The spotted owl loses its habitat when old-growth forests are cut down.

297. (E) Areas that are evaluated for lumber are categorized based upon their site quality.

298. (D) Rotational grazing is similar to the grazing done by wild herds and is brief, intense, and better for the land than long-term grazing. The benefits come from heavy fertilizing by the herds and from the competing weeds getting eaten.

299. (A) (i) Clear-cutting involves removing all trees in a large designated area and then burning whatever is left. Strip cutting involves clear-cuts in smaller areas in thin strips that reseed themselves. Selective cutting involves selecting out certain species from a diverse forest and cutting the mature trees, leaving the smaller trees intact. Shelterwood cutting involves removing the lesser-quality trees first, leaving the healthier trees standing. A few trees get harvested, and then the rest of the mature trees get harvested after all the seedlings have matured.

(ii) The drawbacks of clear-cuts involve increasing runoff, decreasing nutrients in food left for other species, unpleasant appearance of clear-cut areas, soil erosion, fragmentation of wildlife habitat, and loss of species. For strip cutting, the drawbacks are few but may include loss of revenue due to limited areas that can be cut. In the case of selective cutting, it takes more time to harvest fewer trees, and it also removes the best trees, so the forest does not get reseeded with the best trees. In addition, the use of roads and heavy equipment can damage the forest. For shelterwood cutting, the biggest drawback is that it costs the most.

(iii) The benefits of clear-cutting are that nutrients get cycled from burning, and it is cost-efficient. For strip cutting, the benefits include reduced erosion and less impact on the aesthetics of the environment. In selective cutting, the forest appearance is preserved, as is the variety of species. In shelterwood cutting, the aesthetics and habitat are preserved, and erosion is minimized.

(B) Sustainable forestry ensures that the forests provide necessary timber while protecting the forest enough to guarantee that it will be able to provide timber in the future. The forest can be protected this way by (1) using fewer timber products; (2) managing the forest in a holistic manner that accounts for all of the components of the ecosystem, not just the trees; (3) setting aside forestland as a preserve; and (4) replanting trees.

(C) One piece of legislation that protects forests in the United States is the Wilderness Act of 1964, which designates wilderness areas and makes rules about people's activities in those areas. Another piece of legislation is the Federal Land Policy and Management Act of 1976, which sets aside wilderness areas in government-owned areas of forest.

300. (A) Forest fires, although sometimes destructive, can provide some benefits to the forest ecosystem. First, smaller fires eliminate dead branches and leaf litter that can contribute to larger, more destructive fires. Second, some species of trees and other plants require fire for their seeds or cones to germinate. Third, fires can prevent diseases from spreading from tree to tree.

(B) Prescribed fires are those that are intentionally set by trained firefighters in order to eliminate some of the dead tree material on the ground and prevent larger forest fires that would entirely wipe out the forest. In addition, fire is necessary for revitalizing the forest because some of the plants and trees require fire to germinate.

(C) Some of the measures that can protect a home from forest fires include removing branches that hang over your house, keeping leaves and other debris raked up and disposed of properly, mowing grass and keeping it watered, pruning shrubs near the house, and cleaning rain gutters of dried debris.

301. (A) Traditional herding involves letting cows roam freely over grazing land. If their numbers are not too great, this benefits the area because manure acts as a fertilizer, and the grazing stimulates plant growth. This practice is a problem only if the number of cows is too high and the plants are overgrazed to the point that they don't regenerate. Industrial grazing is the same in the beginning. The cattle roam and graze freely, hurting only the environment if the numbers are too large. The major problems occur after that, when the cattle are sent to feedlots. Manure becomes a source of water pollution because of its volume. In addition, gases that are produced by the manure contribute to global warming, the soil may become damaged with salts and trace elements, and the spread of pathogens to humans becomes more viable.

(B) When nonnative grazing animals are introduced to an area of rangeland, the native vegetation may be overgrazed until it's gone because the ecosystem is not used to this species. In addition, if there are other grazing animals on the land, they may have to compete with the new species for food, and their numbers may decrease, or the native species may even be eliminated.

(C) To manage the farm sustainably, it is important to follow these principles of rangeland management:

1. Limit the number of cattle so that the land can regenerate.
2. Rotate the areas that are grazed so their fertility can be restored.
3. Distribute the cattle in different areas and fence off overused areas.
4. Restore the land that has been used by reseeding and fertilizing the vegetation.
5. Improve the land with controlled burns to restore fertility and rid land of unwanted plants.

Chapter 10: Land Use

302. (D) All of the other choices are "push" factors affecting immigration.

303. (C) Degree of urbanization is calculated by dividing the city-dwelling population by the total population, then multiplying by 100% to obtain a percentage figure: (20,000 / 1,000,000) × 100% = 2%.

304. (B) Since the total population remains constant, first calculate the change in the urban population: (100 – 52) – (100 – 55) = (48 urban) – (45 urban) = 3-person increase. Next, calculate the urban growth rate by dividing the change by the city's initial population—i.e., the population that it grew *from*—then multiply by 100% to obtain a percentage figure: (3 / 45) × 100% = 6.7%.

305. (A) Developing countries have the highest current rate of urban growth.

306. (C) The overall migration trend in the United States is headed toward the country's Southwest.

307. (D) Urban sprawl causes an increase in water runoff from increased impervious cover, such as parking lots and streets.

308. (A) The U.S. government provided the first three inducements, but it does not typically fund whole suburban developments, and certainly not on a large regional scale.

309. (C) In rural areas, medical providers are few and far between compared to urban areas.

310. (C) Cities have all of the effects indicated except significantly decreasing cloud formation. In fact, particulate matter that cities release into the atmosphere can encourage nearby cloud formation.

311. (A) Rail transportation options operate on their own schedules.

312. (D) Smart growth is a multipronged strategy using these approaches to develop urban areas around environmental protection and livability.

313. (E) Cluster development concentrates housing and leaves large areas of land open for recreation or other uses.

314. (A) Mixed-use development involves integrating housing, work, and other components of a person's daily life within walking distance.

315. (D) The "fundamental land" policy is quite stringent, with violations that can include the death penalty.

316. (C) Greenbelt is crossed by transit corridors, and it provides space for recreation and plant and animal habitats in cities like Toronto.

317. (E) An ecological footprint is a measure of a person's or population's demand on the environment—the amount of land and ocean necessary to reproduce consumed resources and to dispose of or recycle waste. Each human meal requires agricultural space to grow, for example, and landfill space is often used for disposing of waste.

318. (D) Total impact is calculated by multiplying the other factors by one another, so $I = P \times A \times T$. I would increase fourfold if both P and A were doubled while keeping T the same.

319. (D) Heat and carbon dioxide do not increase local rainfall levels, but the other three factors do contribute to urban flooding.

320. (E) Sewage doesn't do much to increase calcium in oceans and lakes, and calcium doesn't pose a danger to fish and shellfish.

321. (B) Saltwater oxygenating plants do not exist.

322. (B) Agricultural land accounts for around 40% of earth's land area, with forests accounting for about 30% and urban areas making up less than 5%.

323. (D) About 73% of U.S. federal public land is located in Alaska.

324. (E) The Bureau of Land Management oversees areas that often have a commercial use but not a recreational one.

325. (A) The U.S. Fish and Wildlife Service's primary task is to oversee these areas for habitat preservation, although it also allows other uses, such as recreation or oil extraction.

326. (E) The National Wilderness Preservation System includes 660 road-free areas that usually lie within other public land areas. Some are open for recreation.

327. (C) Developers and resource extractors, such as oil companies, have challenged every conservation principle and continue to do so.

328. (A) The tragedy of the commons leads to the degradation of resources by well-meaning people who each do little damage, although the net effect can be disastrous.

329. (C) About half of earth's original forests remain, especially in the taiga of the far Northern Hemisphere.

330. (B) A mineral is composed of several elements in a particular crystalline structure.

331. (D) Placer mining involves using water to wash away unwanted materials.

332. (A) Gillard's goal of increasing job opportunities in outer suburbs will likely increase sprawl around those existing suburbs, unless Australian regional and local governments structure policies to specifically prevent further suburban development.

(B) Each person in a densely constructed urban area can more quickly and easily reach housing, shopping, medical facilities, transportation options, and other needs. In spread-out suburbs, reaching many of these amenities may require a long drive or walk.

(C) Since urban areas with more community services can more efficiently serve a greater number of people, the funding shortfalls in bad economic times have less of an effect on services per person than in spread-out areas.

333. (A) Strip mining is a process in which machines, explosives, or both are employed to remove horizontal layers of earth from a mineral-rich area. Used most often to recover coal and tar sands, strip mining an area uncovers rich seams of the desired material that already lie near the surface. Miners process the removed earth to extract desired materials and discard the remaining waste.

(B) Miners often discard waste earth in nearby rivers and streams that take the waste away. This waste material often contains previously buried poisonous heavy metals such as arsenic, petroleum products from mining machines, and wastewater and sewage from mining facility buildings. Particulate matter often accumulates in nearby streams and can completely bury them, destroying both the streams and the ecosystems that previously relied upon them. At the surface, strip mining completely destroys any ecosystem that previously existed at the site, such as a pine forest or the stream that currently exists at the site of the planned Chuitna project.

(C) The Surface Mining Control and Reclamation Act of 1977 would require PacRim to attempt some environmental restoration only at the end of the Chuitna Coal Project, which might occur after the project's first 25-year phase, at the very soonest.

334. (A) The United States first established national parks to protect undeveloped natural areas and the species that live there. At Yellowstone, this includes the park's incredible geysers, canyons, prairies, forests, and wide diversity of animal and plant species.

(B) Hunters are generally not allowed to kill animals in U.S. national parks. This prohibition has helped the endangered bison and grizzly bears of the West to survive.

(C) The U.S. National Park Service administers and maintains the country's national parks.

(D) Tourism is one of the main threats—even well-meaning visitors erode trails and leave waste behind that degrades the natural environment. The more visitors, the more degradation generally takes place. Surrounding developed areas also threaten national parks through air and water pollution crossing into parkland, as well as the introduction of nonnative species.

Chapter 11: Energy Consumption

335. (A) The "Calorie" used for gauging food energy is equal to one kilocalorie—with a lowercase c—that scientists use for measuring energy in other contexts.

336. (D) The moving of matter is work, whether the matter moved is an apple or the water molecules whose vibrations increase as a pot comes to a boil.

337. (C) The other examples involve matter in motion—an expression of kinetic energy. The water behind the dam has the potential to spill across the land, but it's held in check.

338. (C) Power is work performed divided by a unit of time.

339. (E) This is also known as the first law of thermodynamics.

340. (D) Each step in the process involves some loss of energy as heat, including the coiling of the spring and the flowing of the water down the pipe. Only the sandwich retains all the original energy used in this scenario.

341. (B) Conversion in this case involves sunlight passing through the window, in which some energy is reflected and lost as heat. Inside the room, almost all of the sunlight becomes heat. All the other choices involve several steps in which energy is expended in conversion.

342. (C) Hydrocarbons are composed mostly of hydrogen and carbon.

343. (C) Fossil fuels make up about 86% of U.S. energy usage, while nuclear fuels make up 5% to 10%, wind about 3% to 5%, and geothermal energy less than 1%.

344. (B) Coal, oil, and natural gas are nonrenewable resources.

345. (A) Geothermal power is derived from the heat of the earth's interior.

346. (D) Middle Eastern countries have about 78% of the world's total oil supply beneath them.

347. (A) The Gulf of Mexico contains the United States' largest offshore reserves of oil.

348. (B) The United States uses about one-quarter of the oil produced in the world each year.

349. (A) Saudi Arabian oil can cost as little as $2.50 per barrel. Oil from the United States is very expensive by comparison.

350. (A) In 2003, Canada's bitumen—or tar sands—deposits were reclassified as oil deposits.

351. (C) Conventional and unconventional oil types do not differ significantly in carbon dioxide released upon burning. However, the processing steps necessary for converting some unconventional oil sources into usable forms can release huge amounts of carbon dioxide on their own.

352. (C) Natural gas "floats" above petroleum deposits in the earth, and it contains all these gases, with methane in the largest amount.

353. (D) Like all fossil fuels, burning natural gas results in release of carbon dioxide.

354. (B) Solar energy is renewable and will last as long as the sun—assuming unobstructed skies. Natural gas supplies are projected to last between 60 and 125 years, while conventional oil supplies are projected to last 30 to 60 more years.

355. (B) Poisonous hydrogen sulfide gas, H_2S, is often found in natural gas and must be removed.

356. (C) Methane hydrate is a "frozen" type of unconventional natural gas at the bottom of the ocean.

357. (A) Contour strip mining is used to remove coal from hilly and mountainous terrain.

358. (C) Coal makes up most of the world's buried fossil fuels.

359. (D) Russia has about 31% of the world's natural gas supplies.

360. (C) Natural gas releases less carbon dioxide per unit of energy than any other common fossil fuel.

361. (A) Due to recent improvements in a process called *hydraulic fracturing* or "fracking," which converts shale into natural gas, production in the United States will likely increase until at least 2035.

362. (A) Natural gas is transported in special pressurized pipelines or in containers once it has been cooled to a low enough temperature to liquefy.

363. (C) Anthracite has the highest energy content per unit of mass, while lignite has the lowest.

364. (D) The United States has the largest coal reserves of any country.

365. (D) Crude oil and natural gas are typically found within the same dome formation, with natural gas floating above the oil reservoirs within the deposits.

366. (A) Peat is a fossil fuel, since it consists of ancient, compressed plant matter that cannot be quickly replenished. Biomass is recently produced plant matter and is a renewable resource.

(B) As a lower-quality, dirtier-burning coal, bituminous is composed of a lower proportion of carbon by weight than anthracite. But it has a much higher carbon content than peat, since it has been compressed and chemically converted into coal from ancient peat. (Bituminous coal's exact carbon content varies from 77% to 87%.)

(C) Most likely not. The ancient Irish lake completely filled with plant matter long ago, and it no longer maintains a thin layer of water that would support plant growth and accumulation of dead plant matter, as well as the occasional addition of a sedimentary layer. Second, human interference has drained the bog and removed most of the peat.

367. (A) Hydraulic fracturing involves forcing high-pressure fluid—a liquid or gas—into the rock at the bottom of a well. This fluid forces the nearby rock to crack, releasing pockets of trapped natural gas that can be removed through the well.

(B) More. The production of natural gas by hydraulic fracturing releases much more methane into the atmosphere than the production of conventional natural gas. The two gases produce the same amount of carbon dioxide when burned, however.

(C) At an average of 5 to 250 meters deep, water wells are typically not nearly as deep as wells for producing natural gas from shale, which are usually 1,500 to 6,500 meters deep. So hydraulic fracturing fluid or gas must flow through a great deal of rock to get into water supplies, and it may do so through existing crevices and pores, but the flow of fluids through deep underground rock is not well understood.

368. (A) Some energy is lost as heat at each and every processing step involved in changing coal from one form to another.

(B) The price of petroleum-based fuels would have to rise above that of coal-derived liquid fuels, including the price of infrastructure needed to produce the new fuels. This can happen due to petroleum's becoming more expensive to produce or as a result of its becoming more difficult to obtain—for example, due to an embargo by oil-producing nations. Also, the price of competing fuels, such as biomass-based ethanol, would probably also have to be very high in order for liquid-coal fuels to become a competitive option.

(C) Unburned fuels such as gasoline are highly toxic, and many contain carcinogens. Incomplete combustion can also produce dangerous gases such as carbon monoxide as well as environmentally damaging products, including nitrous oxide.

(D) Coal-derived fuel would still release about as much carbon dioxide as petroleum-based fuel when it is burned, in addition to the considerable land-use, pollution, and health hazards involved in coal mining.

Chapter 12: Nuclear Energy

369. (B) Uranium-238 is nonfissionable, and it makes up about 97% of the uranium in reactor fuel.

370. (B) The splitting of an atomic nucleus is called *fission*.

371. (A) Coolant transfers heat outside the core, where it usually drives turbines to generate electricity.

372. (C) Radioactive decay in the core causes nearby materials to become radioactive themselves, and in the process, it changes the constituent elements of these materials, making them weaker and more likely to crack, break, and fall apart.

373. (E) Uranium-235 contains 235 protons and neutrons altogether—92 protons and 143 neutrons.

374. (D) Alpha particles are composed of protons and (often) neutrons, while beta particles are actually high-energy electrons, which are comparatively low-mass. Gamma rays are high-energy photons—packets of light—and have no mass at all.

375. (C) Two atoms of the same element—one radioactive and the other not—will contain different numbers of neutrons but the same number of protons, electrons, and all other particles, as well as the same chemical properties.

376. (B) Calculate this by first finding the number of half-lives that have passed in 120 years: 120 years / 40 years per half-life = 3 half-lives. Now find out how much material is unchanged after three half-lives by dividing the mass of material in half sequentially three times—once for each half-life: 50 g / 2 = 25 g; 25 g / 2 = 12.5 g; 12.5 g / 2 = 6.25 g. The mass 6.25 g is *unchanged*, while the question asks for how much of the material has decayed after 120 years, so subtract the unchanged mass from the total starting mass: 50 g – 6.25 g = 43.75 g.

377. (D) Radioactive particles release ionizing radiation, which consists of very fast-moving particles that slice important biological molecules into pieces, including DNA.

378. (D) A breeder reactor produces fissionable fuel by adding neutrons to nonfissionable material. A breeder reactor can convert uranium-238 into plutonium-239, for example.

379. (D) Spent nuclear fuel rods are not radioactive enough to create a nuclear explosion, but they are radioactive enough to heat up well beyond the flash point of their constituent materials. If this happens, the resulting fire will release radioactive gases and dust into the environment.

380. (A) Only the "closed" nuclear fuel cycle involves reprocessing spent fuel into usable fuel.

381. (C) High-level radioactive waste that has been reprocessed to remove plutonium-239 must generally be stored for at least 10,000 years for its dangerous levels of radioactivity to subside to safe levels. Unreprocessed waste containing plutonium-239 must be stored for hundreds of thousands of years, with 240,000 years as a commonly cited figure.

382. (E) The United States has not yet begun storing its high-level radioactive waste in long-term facilities, and the one major candidate area at Yucca Mountain has not yet been built, partly due to federal budgetary difficulties. These wastes are mostly stored on-site at nuclear reactors.

383. (C) Earth has no transuranium elements of its own. Natural occurrences of these elements, such as plutonium-239, have all undergone radioactive decay since earth's formation—if they were even present then. People produce these elements by means of nuclear physics, including bombarding heavy-element atoms with neutrons in breeder reactors.

384. (D) Reprocessing of nuclear waste releases some carbon dioxide, but not much compared with energy generated by burning fossil fuels. Normal nuclear power plant operations release very little air pollution.

385. (B) Elements of the nuclear fuel cycle each deal directly with radioactive materials. Decommissioning of power plants, mining ore, and temporary storage of radioactive waste are three components of the *open* nuclear fuel cycle. Reprocessing nuclear waste is an element of the *closed* cycle.

386. (A) Most reactors are light-water reactors, which use a coolant—usually water—to circulate within the core and transport heat out of the core, where it heats water that runs a turbine to generate electricity.

387. (A) The U.S. Nuclear Regulatory Commission oversees nuclear power facilities and the fuel that they use.

388. (B) The Chernobyl reactor was in Ukraine, when it was a part of the Soviet Union.

389. (D) Coal plants are very dirty, but they are relatively simple compared to nuclear power plants, and even with pollution-reducing technology, they require fewer safety measures and less maintenance.

390. (C) With about 20 operational nuclear power plants, the Midwestern states have more nuclear power plants than the other listed regions combined.

391. (B) The Yucca Mountain site is located in the state of Nevada, near the California border.

392. (A) Both widely used uranium-enrichment methods—gaseous diffusion and the gas centrifuge—use uranium hexafluoride.

393. (C) Yellowcake is composed of uranium and oxygen, with the chemical formula U_3O_8.

394. (C) Only nuclear power plants use control rods to regulate the rate of energy production. Control rods block neutrons from passing between fuel rods, and partly removing a control rod will accelerate the rate of nuclear fission within the reactor.

395. (D) In the closed nuclear fuel cycle, decommissioned reactors at the end of their usefulness are just as radioactive as reactors in the open nuclear fuel cycle, so their constituent materials are treated as nuclear waste and usually buried.

396. (B) The Fukushima Daiichi plant experienced a partial meltdown as fuel rods heated up in an uncontrolled fission reaction.

397. (D) Scientists are constantly working on improving fusion reactor designs, but so far no one has made a reactor that produces more power from fuel than is expended during the process.

398. (B) Most reactor designs use the hydrogen isotopes tritium, which has one proton and two neutrons, and deuterium, which has one proton and one neutron.

399. (A) The rods had probably already cracked enough to contaminate the water, and they would probably have continued to contaminate the water further, creating more waste for the plant to dispose of. With even less water circulation, the rods might have caused enough steam to build up to trigger a steam explosion, which could damage the containment pool. If the rods became uncovered, they could heat up enough to catch on fire, releasing radioactive gases into the air.

(B) Allowing fission fuel pellets to gather too close to one another could cause an uncontrolled fission reaction, which could very quickly cause a steam explosion and a fire, releasing radioactive gases into the air.

(C) Boron can absorb neutrons that hit its atomic nuclei. This would decrease the number of neutrons emitted by uranium fuel pellets that collide with the atomic nuclei of uranium elsewhere nearby. This would slow the overall rate of nuclear fission and cool the fuel and its surroundings.

400. (A) Deuterium and tritium are two isotopes of hydrogen. Where normal hydrogen has only a proton in each atomic nucleus, a deuterium atom contains one proton and one neutron in each nucleus, and a tritium atom contains one proton and two neutrons.

(B) The plasma is very, very hot and would melt any physical object touching it. This would probably also cause some of the plasma to cool enough to become merely hot hydrogen gas.

(C) Like most types of nuclear reactors, this General Fusion prototype will use heat—in this case, from hot liquid metal—to turn water into steam, which will drive electricity-generating turbines.

401. (A) Alpha radiation is the high-speed movement of alpha particles—helium nuclei—each of which consists of two protons and two neutrons. Protons have a positive electrical charge, while neutrons have no charge.

(B) Fast-moving electrons by themselves are called *beta radiation*. Electrons always have a negative electrical charge.

(C) Gamma radiation is not repelled or attracted by negative or positive charges. It is composed of high-energy photons—packets of light—which are not electrically charged.

Chapter 13: Alternative and Renewable Energies

402. (B) The two methods usually used to create hydrogen gas are the electrolysis of water and the breakdown of fossil-fuel hydrocarbons. The first requires electricity or heat from a primary source to split molecules of water into hydrogen and oxygen gases, while the second method requires heat to make steam that converts hydrocarbons—another primary energy source—into hydrogen and carbon dioxide gases.

403. (E) *Hydroelectric power* is the general term for electricity generated by damming the flow of a river and using the water to turn turbines.

404. (C) Dams often have a negative impact on the environment, including flooding land, which destroys habitat, and decreasing the flow of fertilizer—in the form of silt—below a dam.

405. (A) Photovoltaic cells produce electrical current when exposed to light.

406. (A) D. L. Staebler and Christopher Wronski discovered that amorphous silicon solar cells lose efficiency when exposed to intense light. The drop in efficiency is usually between 10% and 30% for outdoor solar cells, depending on materials and construction.

407. (D) Hydroelectric power from dammed ocean coves comes from the rising and falling tides, which are themselves driven by the moon's orbit around the earth and the earth's own rotation. Geothermal heat is generated partly by this same tidal stress between the moon's orbit and the earth's rotation, as well as energy released by radioactive elements in its interior. Also, the earth's interior retains some of its heat from its formation.

408. (A) About two-thirds of the energy released by burning ethanol must be consumed in running the agricultural and processing machinery required for making the fuel.

409. (E) Wind energy return on energy invested ranges from about 30:1 to about 7:1, with an average of around 18:1. No other alternative energy sources can match that average.

410. (C) Biomass is burned in furnaces, stoves, and fires, not automobile engines.

411. (C) In the United States, ethanol is primarily produced from corn, and if mass ethanol production skyrockets, corn production for fuel could compete with corn production for food, reducing supply and increasing corn prices.

412. (D) Some U.S. home owners have opted to pipe water underground, where temperatures usually hover at around 10°C (or about 50°F), and back. In the winter in cold regions, this water requires less heat to become hot than water straight from the tap. In the summer, the water can be run through the small pipes of a heat exchanger, which blows indoor air across them to remove heat.

413. (B) Wind turbine power output is proportional to the area swept by its blades. Doubling the swept area doubles the energy a turbine can capture.

414. (D) Along with other electricity-generating devices that rely on rotational motion, wind turbines generate alternating current.

415. (B) Biomass can rely on vegetation and other organic matter that is produced in an unsustainable way, for example through the use of fossil fuel–reliant machines, or through overharvesting. Hydrogen can be produced from unsustainable fossil fuel–derived hydrocarbons. And ethanol can be produced through the use of unsustainable fossil fuel–reliant machines or overharvesting.

416. (B) On the molecular level, the major structural difference between methanol and ethanol is that methanol (CH_3OH) has one carbon atom surrounded by hydrogen, while ethanol (C_2H_5OH) has two carbon atoms surrounded by hydrogen.

417. (B) Also known as a "run-of-river plant," a diversion plant's intake involves only part of a river's total water flow.

418. (B) Photovoltaic cells can capture sunlight, causing electrons to move between the cells' semiconducting layers, creating direct-current electricity. Ethanol can be burned in a piston-driven engine similar to a conventional gasoline engine, and the rotational energy can be captured with magnets to generate electricity. While the second half of an ethanol-based electrical generator is similar to a turbine—rotational energy is captured to create electricity—the generator does not capture the flow of a fluid through the device to create rotation, as a turbine does.

419. (C) Burning hydrogen releases only heat and water vapor.

420. (A) Burning ethanol releases mostly carbon dioxide and water—two molecules of carbon dioxide and three molecules of water per ethanol molecule burned. However, impurities in the air and fuel, as well as incomplete combustion, can create other products, such as undesirable pollutants carbon monoxide and nitric oxide. These are never the majority products, though.

421. (A) Under controlled conditions, bacteria can convert biomass, such as wood, leaves, and other plant parts, to methane, methanol, or ethanol.

422. (E) The United States and Brazil produce and consume between 80% and 90% of the ethanol fuel in the world.

423. (D) The more steps between an energy source's original form and its final use, the more energy is lost along the way. Here energy is lost burning the biomass, heating water to steam, turning the turbine for electricity, making hydrogen from water, and burning hydrogen in an automobile. Other choices use some of these same steps but fewer of them.

424. (C) In the United States, a 10% liquid ethanol, 90% gasoline mixture is used in ordinary automobiles.

425. (B) It takes nearly twice as much ethanol as gasoline to move an automobile the same distance.

426. (B) Only organic materials are useful as biomass fuel.

427. (C) Reykjavik, Iceland, is the home of the world's largest geothermal district heating utility.

428. (B) Depending on local conditions, underground heat can be depleted for a period of time that is significant to people, such as a period of months or years. To slow heat depletion, some facilities that use geothermal steam will pump used hot water back into underground reservoirs.

429. (B) Renewable energy certificates, or green tags, can be bought, sold, and bartered in their own market apart from the actual physical electricity produced by a renewable energy supplier, such as a wind farm.

430. (E) With the extremely low melting temperature of –259.14°C—about 14 K—hydrogen would require too much energy to freeze for storage.

431. (C) Biomass produces more electricity than all other alternative energy sources combined.

432. (A) PS10 is a thermal solar plant rather than a photovoltaic plant because it uses the sun's light to make heat that generates electricity.

(B) Although its maintenance and operation costs haven't yet been demonstrated, PS10's advantage is that it doesn't need a constant supply of fuel, and while it has a steam turbine—like coal-fired plants—it doesn't have much of the other complicated machinery, such as systems for removing sulfur and heavy metals from exhaust. There is also no need to dispose of coal ash or much of any other type of waste from PS10.

(C) PS10 does not emit any carbon dioxide, sulfur, nitrogenous compounds, heavy metals, or any other pollutant. It also does not require dangerous and environmentally harmful mining to supply its energy. However, PS10 requires a fairly large piece of land: (624 mirrors) × (120 square meters per mirror) = 74,880 square meters minimum space.

433. (A) Hoover Dam is an impoundment dam, since it retains water in a reservoir, letting a controlled, high-pressure stream out to do work. In the case of Hoover Dam, the water stream runs turbines that generate electricity.

(B) Yes. Lake Mead fills with water from upstream rainfall and melting snow, which are replenished by natural processes.

(C) Lake Mead consumed a large area of land that once featured plants and animals belonging to the local desert ecosystem. Also, the dam prevents the flow of nutrients to downriver areas, and it interferes with local aquatic ecosystems, such as those of river fish.

434. (A) Less. Rain and snowfall have almost no contribution to the Rance power plant's electricity generation, because tidal waters would continue to rise and fall without any regional rain or snowfall at all.

(B) The sources of tidal energy are the earth's rotation on its axis and the gravitational forces from the moon and the sun.

(C) The enclosed estuary can fill wi[illegible] silt—as has happened with the Rance plant to some degree—affecting local p[illegible] animals by changing their habitat. Also, species that once lived in the estua[illegible] nearly completely cut off from the ocean, which is a problem for plants and anim[illegible] that depend on access to its waters.

Chapter 14: Pollution Types

435. (D) Smog levels are highest in summer because the reactions that form smog are affected by sunlight, and the air is relatively hot and slow moving in summer.

436. (A) The CAIR set limits on emissions of sulfur dioxide, nitrogen oxides, and particulates.

437. (B) Agriculture pollutes more than half of the streams in the United States, mostly by fertilizer and pesticide runoff.

438. (C) Municipal water pollution comes from residential and commercial wastewater, which comes from sources such as sewage.

439. (E) Dinoflagellate (or red tide) blooms are caused by storm water runoff and can cause fish and mammal deaths and threaten human health.

440. (A) The amount of oxygen available in water is affected by temperature (cold is best), how fast and how big the stream is (bigger and faster is better), and how many plants are producing oxygen (more plants are better). Another factor is how many species are using up the oxygen that is produced.

441. (D) Chlorination is used in most large cities because chlorine stays in the water after leaving the treatment plant as opposed to other methods that disinfect at the treatment plant but leave open the possibility that the water gets contaminated again after it leaves.

442. (C) Water from waste treatment plants is considered a point source because it is released directly into the water supply, whereas the other sources of contamination indirectly lead to the water supply.

443. (B) Phosphorus occurs in low concentrations naturally but is found in high concentrations near populated areas. Excessive phosphate levels lead to eutrophication, in which the oxygen content of water gets depleted. Fish are among the organisms that can die as a result.

444. (E) Acid rain affects trees in many ways: damaging the coating on leaves disrupts evaporation and gas exchange, leaching nutrients from soil makes them unavailable to the tree, and toxic metal buildup combined with acid rain stunts growth.

445. (E) The Air Quality Accord established a permanent limit on how many tons of acid emissions each country could emit. Sources of acid emissions include fuels with high sulfur content, sulfur dioxide from smokestacks, and automobiles and trucks without catalytic converters.

446. (C) While radioactive waste is associated with nuclear fission, several industries and products are connected with the production of radioactive waste, including power plants, industry, mining, and medical tracers.

447. (A) Earth Day was established on April 22, 1970, to raise public awareness about environmental issues including pollution.

448. (E) Chlorofluorocarbons break down in sunlight, where chlorine-free radicals combine with ozone, breaking it down. They also account for approximately one-quarter of all greenhouse gases that contribute to global warming.

449. (D) Methyl mercury is the by-product that causes health problems.

450. (B) This act of Congress was signed by almost 30 nations.

451. (C) An example of a synergistic reaction occurs between chlorine and some organics to form chloramines, which are more toxic than just the chlorine by itself. The health effects of chloramines can include irritating the respiratory system, and it is unknown if they are possibly carcinogenic as well.

452. (A) Dioxins cause several health problems, which is especially problematic since herbicides are produced in mass amounts and are the most commonly used pesticide.

453. (E) Particulates can cause breathing problems and on a large enough scale can contribute to global warming and ozone depletion.

454. (B) Burning of oil and gas produces all of these chemicals.

455. (B) The Toxic Substances Control Act regulates chemicals in many ways. First, it requires that the EPA be notified before a chemical is made. Second, it controls substances that could endanger people or the environment. Third, chemicals that were made prior to the 1976 law have to be tested for safety as well.

456. (C) Teratogens are substances that cause birth defects. Although some of these also cause cancer, they are all known to cause birth defects.

457. (A) Anthropogenic pollution is caused directly by humans.

458. (D) Noise level or loudness is measured in decibels (dB), with high levels over 100 dB causing hearing damage.

459. (E) Formaldehyde is often found in carpeting, furniture upholstery, plywood, glues, and other products found in homes. It can cause cancer or problems in a person's respiratory tract.

460. (D) Tradable permits encourage companies to lower emission rates so they can sell the rest of their quota for money.

461. (B) Also known as *environmental tobacco smoke*, secondhand smoke contains carbon monoxide, sulfur dioxide, nitrogen dioxide, and particulates.

462. (A) Ozone is produced as a secondary pollutant from the combination of sunlight and automobile emissions.

463. (C) Lichens absorb almost everything from the air, including pollution.

464. (E) Radon is a radioactive gas that comes from the earth.

465. (A) The LD_{50} is the dose of a chemical at which half of the animals exposed to it die, also known as the lethal dose for 50% of the population. The TD_{50} is the dose of a chemical that is toxic to half of the population, causing some kind of visible negative effect. These measurements are used to determine how toxic a chemical is to a population and whether or not its benefits outweigh its risks.

(B) When the Clean Water Act has no threshold for certain pollutants, it means that no amount of those pollutants is considered acceptable. This assumes that there is no safe level of pollutant that can be tolerated in water. It is simply a zero-tolerance policy.

(C) An ecological gradient near a smokestack that emits SO_2 would show the most loss of plant life where the pollutant concentration is strongest. As the distance from the smokestack increased, the vegetation would be healthier and healthier. The only exception to this rule occurs if the emissions cause acid rain, the area in which pollutants have an effect can be increased.

466. (A) Potential sources of mercury include power plants fueled by coal, mining, and industrial manufacturing.

(B) Mercury enters the human body primarily by drinking water and eating fish.

(C) Human health effects of mercury include tremors, loss of hair and teeth, numbness, loss of neurological functions, and death.

(D) Mercury enters water and is changed to two different types of mercury by bacteria in the water. Methyl mercury, one of these forms, stays in the bottom sediment and is released over time. A small fish feeds on plants that contain mercury. Then a medium-sized fish eats that fish and others of similar size and accumulates more mercury in its system. The largest fish, such as tuna, eat many of the medium-sized fish and have the greatest mercury concentration of all the fish, making them one of the most dangerous for people to consume frequently.

467. (A) Lead poisoning disproportionately affects those in poverty because they tend to live in urban areas where lead was persistent in car exhaust. The lead was then distributed to the soils in the area and has persisted for many years. Also, older houses, which tend to be in more impoverished areas, may have leaded paint that children can be exposed to if they eat it or if it is present in dust or fumes.

(B) Lead poisoning can be prevented by reducing emissions in power plants and other industrial processes, as well as from car gasoline. Also, by replacing lead pipes, it can be further eliminated from drinking water. Homes can be tested for lead if they are older, and the paint can be removed, or objects with lead paint can be removed from the home.

(C) People who live in nonindustrialized countries are still exposed to lead because it is in the atmosphere that was polluted by industrialized countries and therefore contaminates their air and water as well.

Chapter 15: Global Change and Economics

468. (B) Climate describes the conditions in the atmosphere that are typical for that area for many years. Weather also describes atmospheric conditions but just in the short term.

469. (A) Carbon dioxide levels can be determined by looking at the air bubbles trapped in ice cores, with changes noted at different levels in the core.

470. (C) About a third of the sunlight reaching earth is reflected back into space, while the rest is absorbed by everything on earth and then gradually goes into the atmosphere.

471. (D) Both deforestation and burning of fossil fuels increase the amount of CO_2 (a greenhouse gas). That increase, as well as the increase in other greenhouse gases, increases the amount of heat radiated back to earth.

472. (E) Aerosols are made up of chlorofluorocarbons. The other choices are all sources of methane.

473. (B) Chlorofluorocarbons (CFCs) are the only greenhouse gases that are solely man-made.

474. (C) Photodissociation is the process by which things break down in sunlight. All three compounds are dangerous when broken down by sunlight because free radicals are formed that then react with ozone, breaking it down.

475. (A) The ozone layer absorbs UV radiation, and without its protection, there would be more skin cancer, cataracts, and damage to plants and animals.

476. (C) Warming temperatures both melt the glaciers and cause an expansion of water in the ocean, leading to rising sea levels.

477. (E) El Niño creates warmer ocean temperatures, which contribute heat to the atmosphere.

478. (D) Insects that live in tropical areas and carry disease might benefit from global warming because the size of their potential habitat would increase.

479. (B) Oceans, as well as trees in forests, absorb carbon dioxide, acting as a "sink," or a place to store carbon dioxide. When any of these is destroyed, the carbon dioxide it was storing gets released.

480. (C) Even though the United States has only 5% of the world's population, it produces 20% of the world's carbon dioxide emissions.

481. (A) Natural gas releases the smallest amount of carbon. Coal releases the greatest amount of carbon.

482. (B) Polar amplification occurs when the greenhouse effect is amplified by more heat being absorbed by the ocean instead of reflected away by ice, especially in the polar regions.

483. (D) Recycled products often require much less energy to reproduce compared to production of new products, and aluminum is more expensive to produce new compared to glass or steel.

484. (E) The lumber company would plan on direct costs, such as the cost of machinery to cut the lumber and the price of fuel to operate the machines. However, other costs, such as loss of tourism dollars and loss of medicinal plants for pharmaceuticals, are known as *externalities*, or *indirect costs*, and are sometimes not taken into consideration by companies when pursuing their business.

485. (E) The United States has a free market economy, where consumer decisions drive supply and demand. Countries that are communist and some that are controlled by dictators often have a command economy in which the government makes all of the decisions. This often benefits those in power, and the majority of the people do not profit or get the goods they need.

486. (C) Policy instruments can be used to urge everyone to follow through on environmental goals that affect everyone.

487. (D) Repercussion costs occur as a consequence of environmental damage, when a company loses money from bad public image, boycott, or other public reaction.

488. (C) The Index of Sustainable Economic Welfare (ISEW) is a measure of economic progress that takes into account the costs of pollution and other environmental costs, as well as other hazards detrimental to public safety and health (such as highway safety).

489. (A) Marginal cost is used to calculate how economically feasible it is to clean up or control pollution.

490. (B) Air pollution has a higher risk of death than any of the other activities listed due to its contribution to heart and lung disease, asthma, emphysema, and other respiratory functions.

491. (C) Western nations can help developing nations by providing tools that don't require extensive infrastructure and tools and that do not harm the environment in other ways.

492. (D) Glacial ice cores preserve samples of atmospheric gases.

493. (E) A reaction in which the product can be used over and over again in future reactions is known as a *catalytic chain reaction*, as is the case with chlorine from the breakdown of CFCs.

494. (B) Nitrogen oxides come from combustion and are efficient at trapping heat in the atmosphere.

495. (E) Industrialization has led to an increase of about 30% in CO_2 concentration as compared to 250 years ago.

496. (A) How long a company is willing to wait for greater profits in the future is known as its *time preference.*

497. (D) The beauty offered by natural places as well as the indirect services that some living things provide to humans are often overlooked or unnoticed, making it difficult to incorporate those aspects into how much a particular environment is valued and preserved.

498. (A) The amount of energy coming from the sun changes over time, with the warmer periods on earth correlating to times in which more energy was being radiated from the sun, and the Ice Ages correlating to times when there was less solar radiation. Although there is a correlation, the effect is small compared to that of greenhouse gases, especially those generated by human activities. Volcanoes contribute to global cooling temporarily because the volcanic ash interferes with solar radiation. Greenhouse gases such as carbon dioxide absorb solar radiation that is being emitted toward space, increasing the temperatures here on earth. Weather system changes such as El Niño occur approximately every five years and last just over a year. Their occurrence is expected to be more frequent as the earth warms. They cause the ocean to become a heat source, temporarily increasing temperatures in the atmosphere.

(B) Negative feedback will help keep global warming in check. An example of negative feedback in global warming is when the warmer air caused by global warming then creates more precipitation in northern latitudes. This can increase the amount of snow and ice in the area, which will reflect more of the incoming energy from the sun back to space, leading to cooling.

(C) Positive feedback will make global warming an even bigger problem. An example of positive feedback would be that warming temperatures would cause snow and glaciers to melt, which means less energy is reflected back into space and is instead absorbed by the atmosphere, leading to even higher temperatures.

499. Dear family,
I know there has been a lot of controversy surrounding global warming. There are a variety of opinions and possible scenarios that we might see in the future. I am writing to you out of concern for the well-being of each family member. Uncle Joe, I know you are already struggling to make money on your farm there in Iowa. Well, we will probably expect to see more droughts, hotter summers that will lessen the corn harvest, and more competition from farmland that becomes available in

Canada. Aunt Susanna, I am concerned for your safety after the last flood. I know you cannot get your house insured now for flooding, and now that you are getting older, I wonder how you will evacuate and rebuild in the event of the next hurricane. Grandma Betty, we need to make some changes after a few of your neighbors were hospitalized for West Nile virus last summer. We need to write to your local legislators and insist that government intervention is stepped up in mosquito control and eradication. We also need to insist on a cap of greenhouse gas emissions for the state of New Jersey, as well as higher mileage requirements for cars. Since you are getting older, West Nile virus is more of a threat to you than to younger people. There is a bill coming up for a vote in Congress this week on reducing greenhouse gas emissions. Please contact your local senator to support this vote because it will help reduce the problems I listed above. Thanks.
Love, Sam

500. **(A)** The cost of preventative measures is represented by curve C. This is because the more the pollution level is controlled, the more expensive it gets.

(B) The costs for cleanup are represented by curve B. This is because the more pollution that is removed, the less cost there is from pollution damage that will need to be cleaned up.

(C) Point A represents the point at which the cost of removing pollution and the benefits that come from removing pollution are equal. The information from point A will tell me how much money to spend to get a reasonable amount of pollution cleaned up. I would not spend beyond the cost at point A for pure economic reasons. However, if the company valued the environment or wanted the image that it valued the environment, it may clean up more than that to include the costs to the habitat, the effects on public health in the region, and the effects on other parts of the economy. These factors are also known as *externalities*.